Concepts and Breeding of Heterosis in Crop Plants

Related Society Publications

Conservation of Crop Germplasm–An International Perspective

Genetic Contributions to Yield Gains of Five Major Crop Plants

International Germplasm Transfer: Past and Present

Use of Plant Introductions in Cultivar Development: Part 1

Use of Plant Introductions in Cultivar Development: Part 2

For information on these titles, please contact the ASA, CSSA, SSSA Headquarters Office; Attn.: Marketing; 677 South Segoe Road; Madison, WI 53711-1086. Phone: (608) 273-8080. Fax: (608) 273-2021.

Concepts and Breeding of Heterosis in Crop Plants

Proceedings of the Plant Breeding Symposium sponsored by the Crop Science Society of America and the American Society of Horticultural Science, 3 November 1996, in Indianapolis, Indiana.

Editors
Kendall R. Lamkey and Jack E. Staub

Organizing Committee
Kendall R. Lamkey, *Co-chair, Crop Science Society of America*
Jack E. Staub, *Co-chair, American Society of Horticultural Science*
Paula J. Brammel-Cox
John W. Dudley
James Nienhuis
Roger E. Freeman

Editor-in-Chief CSSA
Jeffrey J. Volenec

Managing Editor
David M. Kral

Associate Editor
Marian K. Viney

CSSA Special Publication Number 25

Crop Science Society of America
Madison, Wisconsin

1998

Cover design by Patricia Scullion; photograph provided by Dwight Tomes

Copyright © 1998 by the Crop Science Society of America, Inc.

ALL RIGHTS RESERVED UNDER THE U.S. COPYRIGHT ACT OF 1976 (PL. 94-553).

Any and all uses beyond the limitations of the "fair use" provision of the law require written permission from the publisher(s) and/or the author(s); not applicable to contributions prepared by officers or employees of the U.S. Government as part of their official duties.

Crop Science Society of America, Inc.
677 South Segoe Road, Madison, WI 53711 USA

Library of Congress Registration Number: 98-71578

Printed in the United States of America

CONTENTS

Foreword vii
Preface ix
Contributors xi
Conversion Factors for SI and non-SI units xiii

1 From Out of Old Fields Comes All This New Corn:
    An Historical Perspective on Heterosis in Plant Improvement
        I.L. Goldman.. 1

2 Heterosis: Performance Stability, Adaptability to Changing
    Technology, and the Foundation of Agriculture as a Business
        Dwight T. Tomes................................... 13

3 Overview of Heterosis and Heterotic Groups in Agronomic Crops
        Albrecht E. Melchinger and Ramesh K. Gumber........... 29

4 Hybrids in Horticultural Crops
        Jules Janick....................................... 45

5 Apparent Overdominance in Natural Plant Populations
        Jeffry B. Mitton................................... 57

6 Role of Chromosome Blocks in Heterosis and Estimates
    of Dominance and Overdominance
        Edwin T. Bingham.................................. 71

7 Inference of Heterosis and Epistasis in Transposon
    Tagged *Drosophila*
        Andrew G. Clark................................... 89

8 Avoiding Project Bankruptcy While Effectively Employing Markers
    Tom Blake, Steve Larson, Joy Eckhoff,
        and Vladimir Kanazin............................. 99

9 Molecular Analyses and Heterosis in the Vegetables:
    Can We Breed Them Like Maize?
        Michael J. Havey................................. 109

10 Predicting the Performance of Untested Single Crosses:
    Trait and Marker Data
        Rex Bernardo 117

FOREWORD

Heterosis (or hybrid vigor) has been a major factor in the increased production of several important plant species during this century; however, the most dramatic and well-publicized impact on yield and other performance characteristics has probably occurred in maize. From 1933 to 1943 the increased yield of hybrid maize over the traditional open-populated varieties was so evident that approximately 90% of all farmers in the Midwestern portion of the USA switched to planting commercial hybrids.

Two very significant results occurred from the development of hybrid maize in the USA. First was the development of a far superior product for farmers. Second was the rapid increase in the number of researchers working in the private sector to develop improved maize varieties. Prior to 1930, most of the professional researchers, which were few in number, were employed by the public sector. Today, more than 500 researchers in the USA work on maize breeding, and more than 90% of them are employed in the private sector. There is little doubt that the exploitation of heterosis and the growth in number of researchers working on maize were major factors contributing to the increase in average yield in the USA from 1.5 Mg ha^{-1} to more than 7 Mg ha^{-1} during the period 1930 to 1990.

The alarming rate at which the world population continues to grow demands that major cereal production will have to increase by 30 to 40% over current production by the year 2020. One area of research that could prove to be valuable in achieving this increase is the exploitation of heterosis and the development of commercial hybrids in wheat and rice, the two major food cereals. Hybrid rice is now a viable commercial product in countries such as China and India that rely on this plant as their major food cereal. Area devoted to hybrid rice is growing rapidly in each of these countries.

In light of the pressing problem of producing ever-increasing amounts of food on less arable land, this publication on the history and current knowledge of heterosis is most timely. The Crop Science Society of America appreciates very much the hard work of K.R. Lamkey and J.E. Staub in performing the many time-consuming duties associated with editing this book.

Ronald P. Cantrell, *President, 1998*
Crop Science Society of America

PREFACE

The Joint Plant Breeding Symposia Series is sponsored by the Crop Science Society of America (CSSA) and the American Society of Horticultural Science (ASHS) to provide information and increase the awareness of plant breeders and plant geneticists in important areas of applied genetics. The symposia is held every other year and alternately hosted by the two societies. Concepts and breeding of heterosis in Crop Plants, the third joint plant breeding symposia, was held on Sunday, 3 November 1996 in Indianapolis at the ASA–CSSA–SSSA national meetings. The topic of heterosis grew out of discussions between committee members who had organized the previous symposia held in 1994 and members of the Vegetable Breeders Working Group, ASHS. They felt there was a need to bring together agronomic and horticultural scientists to discuss issues related to heterosis and its application to crop species.

The resulting one-day symposium was organized around three general themes: history and perspective, mechanisms, and detection and exploitation. The intent of the history section was to provide a historical overview of heterosis, its role in business, and to present a summary of the utility of heterosis in agronomic and horticultural crops. The intent of the mechanisms of heterosis section was to examine the underlying genetics and biology of heterosis. Likewise, the intent of the detection and exploitation of heterosis section was to examine how heterosis can be manipulated and utilized in practice.

Heterosis dominated the thinking of plant and animal geneticists in the 1940s and 1950s as evidenced by the now classic book entitled Heterosis edited by John W. Gowen and published by Iowa State University Press. In fact, the entire U.S. hybrid maize industry and much of the world maize industry is founded on heterosis. Despite the importance of heterosis in maize and the plethora of research done on heterosis, little is known empirically about the underlying genetic and related physiological basis of heterosis. In fact, the biological basis of heterosis is considered by some scientists to be inadequately defined. The chasm of unanswered questions regarding heterosis is undoubtedly why it still dominates the thinking of many scientists and was the impetus for the manuscripts published in this volume.

Although heterosis and hybrid vigor are often used synonymously, heterosis and hybrids are not necessarily synonymous. To have heterosis you need hybrids, but producing hybrids does not guarantee heterosis. The development of hybrids is being actively pursued in many agronomic and horticultural crops, but exploitation of heterosis is often not the primary reason. For instance, many horticultural crop species fail to demonstrate inbreeding depression and/or heterosis. Although the production of hybrids from inbred lines allows one to fix heterosis, hybrids also provide many other advantages in a crop production system. For instance, hybrids provide uniformity, require farmers to purchase new seed each year, and create a vehicle for the protection of intellectual property.

One issue that emerged from this symposium is that heterosis can be difficult to measure and define, particularly in crops where there are multiple harvest dates and related quality issues. Moreover, the importance of yield and quality can differ among crop species. The information found herein should give breeders, geneticists, and biologists additional insights into the complexity of heterosis. It is clear from the symposium contents that old questions regarding heterosis and its relative importance resurfaced, and new questions were defined in the light of emerging technologies.

Kendall R. Lamkey and Jack E. Staub

CONTRIBUTORS

Rex Bernardo — Research Scientist, Limagrain Genetics, 4805 West Old Church Road, Champaign, IL 61821. Currently Assistant Professor of Maize Genetics and Breeding, Department of Agronomy, 1150 Lilly Hall of Life Sciences, Purdue University, West Lafayette, IN 47907-1150

Edwin T. Bingham — Professor of Agronomy, Agronomy Department, University of Wisconsin, 1575 Linden Drive, Madison, WI 53706

Tom Blake — Professor, Barley Breeding and Genetics, Department of Plant Science, Montana State University, Bozeman, MT 59717

Andrew G. Clark — Professor of Biology, Department of Biology, 208 Mueller Laboratory, Pennsylvania State University, University Park, PA 16802

Joy Eckhoff — Associate Professor of Agronomy, Eastern Agricultural Research Center, Montana Agricultural Experiment Station, P.O. Box 1350, Sidney, MT 59270

I.L. Goldman — Associate Professor of Horticulture, Department of Horticulture, University of Wisconsin, 1575 Linden Drive, Madison, WI 53706

Ramesh K. Gumber — Research Scientist, 350 Institute of Plant Breeding, Seed Science, and Population Genetics, University of Hohenheim, D-70593 Stuttgart, Germany

Michael J. Havey — USDA Research Geneticist and Associate Professor of Horticulture, USDA-ARS, Department of Horticulture, University of Wisconsin, 1575 Linden Drive, Madison, WI 53706

Jules Janick — Professor of Horticulture, Department of Horticulture, 1165 Horticulture Building, Purdue University, West Lafayette, IN 49071-1165

Vladimir Kanazin — Research Associate, Department of Plant Science, Montana State University, Bozeman, MT 59717

Steve Larson — Research Associate, USDA-ARS, National Small Grains Germplasm Collection, Box 307, Aberdeen, Idaho 83210

Albrecht E. Melchinger — Professor of Plant Breeding, 350 Institute of Plant Breeding, Seed Science, and Population Genetics, University of Hohenheim, D-70593 Stuttgart, Germany

Jeffry B. Mitton — Professor of Biology, Department of Environmental, Population, and Organismic Biology, University of Colorado, Campus Box 334, Boulder, CO 80309-0334

Dwight T. Tomes — Senior Scientist, Department of Biotechnology, Pioneer Hi-Bred International, Inc., 7300 NW 62nd Avenue, P.O. Box 1004, Johnston, IA 50131-1004

Conversion Factors for SI and non-SI Units

Conversion Factors for SI and non-SI Units

To convert Column 1 into Column 2, multiply by	Column 1 SI Unit	Column 2 non-SI Units	To convert Column 2 into Column 1, multiply by
Length			
0.621	kilometer, km (10^3 m)	mile, mi	1.609
1.094	meter, m	yard, yd	0.914
3.28	meter, m	foot, ft	0.304
1.0	micrometer, µm (10^{-6} m)	micron, µ	1.0
3.94×10^{-2}	millimeter, mm (10^{-3} m)	inch, in	25.4
10	nanometer, nm (10^{-9} m)	Angstrom, Å	0.1
Area			
2.47	hectare, ha	acre	0.405
247	square kilometer, km² (10^3 m)²	acre	4.05×10^{-3}
0.386	square kilometer, km² (10^3 m)²	square mile, mi²	2.590
2.47×10^{-4}	square meter, m²	acre	4.05×10^3
10.76	square meter, m²	square foot, ft²	9.29×10^{-2}
1.55×10^{-3}	square millimeter, mm² (10^{-3} m)²	square inch, in²	645
Volume			
9.73×10^{-3}	cubic meter, m³	acre-inch	102.8
35.3	cubic meter, m³	cubic foot, ft³	2.83×10^{-2}
6.10×10^4	cubic meter, m³	cubic inch, in³	1.64×10^{-5}
2.84×10^{-2}	liter, L (10^{-3} m³)	bushel, bu	35.24
1.057	liter, L (10^{-3} m³)	quart (liquid), qt	0.946
3.53×10^{-2}	liter, L (10^{-3} m³)	cubic foot, ft³	28.3
0.265	liter, L (10^{-3} m³)	gallon	3.78
33.78	liter, L (10^{-3} m³)	ounce (fluid), oz	2.96×10^{-2}
2.11	liter, L (10^{-3} m³)	pint (fluid), pt	0.473

CONVERSION FACTORS FOR SI AND NON-SI UNITS

To convert Column 1 into Column 2, multiply by	Column 1 SI Unit	Column 2 non-SI Unit	To convert Column 2 into Column 1, multiply by
Mass			
2.20×10^{-3}	gram, g (10^{-3} kg)	pound, lb	454
3.52×10^{-2}	gram, g (10^{-3} kg)	ounce (avdp), oz	28.4
2.205	kilogram, kg	pound, lb	0.454
0.01	kilogram, kg	quintal (metric), q	100
1.10×10^{-3}	kilogram, kg	ton (2000 lb), ton	907
1.102	megagram, Mg (tonne)	ton (U.S.), ton	0.907
1.102	tonne, t	ton (U.S.), ton	0.907
Yield and Rate			
0.893	kilogram per hectare, kg ha^{-1}	pound per acre, lb acre^{-1}	1.12
7.77×10^{-2}	kilogram per cubic meter, kg m^{-3}	pound per bushel, lb bu^{-1}	12.87
1.49×10^{-2}	kilogram per hectare, kg ha^{-1}	bushel per acre, 60 lb	67.19
1.59×10^{-2}	kilogram per hectare, kg ha^{-1}	bushel per acre, 56 lb	62.71
1.86×10^{-2}	kilogram per hectare, kg ha^{-1}	bushel per acre, 48 lb	53.75
0.107	liter per hectare, L ha^{-1}	gallon per acre	9.35
893	tonnes per hectare, t ha^{-1}	pound per acre, lb acre^{-1}	1.12×10^{-3}
893	megagram per hectare, Mg ha^{-1}	pound per acre, lb acre^{-1}	1.12×10^{-3}
0.446	megagram per hectare, Mg ha^{-1}	ton (2000 lb) per acre, ton acre^{-1}	2.24
2.24	meter per second, m s^{-1}	mile per hour	0.447
Specific Surface			
10	square meter per kilogram, m^2 kg^{-1}	square centimeter per gram, cm^2 g^{-1}	0.1
1000	square meter per kilogram, m^2 kg^{-1}	square millimeter per gram, mm^2 g^{-1}	0.001
Pressure			
9.90	megapascal, MPa (10^6 Pa)	atmosphere	0.101
10	megapascal, MPa (10^6 Pa)	bar	0.1
1.00	megagram, per cubic meter, Mg m^{-3}	gram per cubic centimeter, g cm^{-3}	1.00
2.09×10^{-2}	pascal, Pa	pound per square foot, lb ft^{-2}	47.9
1.45×10^{-4}	pascal, Pa	pound per square inch, lb in^{-2}	6.90×10^3

(continued on next page)

Conversion Factors for SI and non-SI Units

To convert Column 1 into Column 2, multiply by	Column 1 SI Unit	Column 2 non-SI Units	To convert Column 2 into Column 1, multiply by
Temperature			
1.00 (K − 273)	Kelvin, K	Celsius, °C	1.00 (°C + 273)
(9/5 °C) + 32	Celsius, °C	Fahrenheit, °F	5/9 (°F − 32)
Energy, Work, Quantity of Heat			
9.52×10^{-4}	joule, J	British thermal unit, Btu	1.05×10^{3}
0.239	joule, J	calorie, cal	4.19
10^{7}	joule, J	erg	10^{-7}
0.735	joule, J	foot-pound	1.36
2.387×10^{-5}	joule per square meter, J m^{-2}	calorie per square centimeter (langley)	4.19×10^{4}
10^{5}	newton, N	dyne	10^{-5}
1.43×10^{-3}	watt per square meter, W m^{-2}	calorie per square centimeter minute (irradiance), cal cm^{-2} min^{-1}	698
Transpiration and Photosynthesis			
3.60×10^{-2}	milligram per square meter second, mg m^{-2} s^{-1}	gram per square decimeter hour, g dm^{-2} h^{-1}	27.8
5.56×10^{-3}	milligram (H$_2$O) per square meter second, mg m^{-2} s^{-1}	micromole (H$_2$O) per square centimeter second, m mol cm^{-2} s^{-1}	180
10^{-4}	milligram per square meter second, mg m^{-2} s^{-1}	milligram per square centimeter second, mg cm^{-2} s^{-1}	10^{4}
35.97	milligram per square meter second, mg m^{-2} s^{-1}	milligram per square decimeter hour, mg dm^{-2} h^{-1}	2.78×10^{-2}
Plane Angle			
57.3	radian, rad	degrees (angle), °	1.75×10^{-2}

CONVERSION FACTORS FOR SI AND NON-SI UNITS

Electrical Conductivity, Electricity, and Magnetism

To convert Column 1 into Column 2, multiply by	Column 1 SI Unit	Column 2 non-SI Unit	To convert Column 2 into Column 1, multiply by
10	siemen per meter, S m^{-1}	millimho per centimeter, mmho cm^{-1}	0.1
10^4	tesla, T	gauss, G	10^{-4}

Water Measurement

9.73 × 10^{-3}	cubic meter, m^3	acre-inches, acre-in	102.8
9.81 × 10^{-3}	cubic meter per hour, m^3 h^{-1}	cubic feet per second, ft^3 s^{-1}	101.9
4.40	cubic meter per hour, m^3 h^{-1}	U.S. gallons per minute, gal min^{-1}	0.227
8.11	hectare-meters, ha-m	acre-feet, acre-ft	0.123
97.28	hectare-meters, ha-m	acre-inches, acre-in	1.03 × 10^{-2}
8.1 × 10^{-2}	hectare-centimeters, ha-cm	acre-feet, acre-ft	12.33

Concentrations

1	centimole per kilogram, cmol kg^{-1}	milliequivalents per 100 grams, meq 100 g^{-1}	1
0.1	gram per kilogram, g kg^{-1}	percent, %	10
1	milligram per kilogram, mg kg^{-1}	parts per million, ppm	1

Radioactivity

2.7 × 10^{-11}	becquerel, Bq	curie, Ci	3.7 × 10^{10}
2.7 × 10^{-2}	becquerel per kilogram, Bq kg^{-1}	picocurie per gram, pCi g^{-1}	37
100	gray, Gy (absorbed dose)	rad, rd	0.01
100	sievert, Sv (equivalent dose)	rem (roentgen equivalent man)	0.01

Plant Nutrient Conversion

	Elemental	Oxide	
2.29	P	P$_2$O$_5$	0.437
1.20	K	K$_2$O	0.830
1.39	Ca	CaO	0.715
1.66	Mg	MgO	0.602

1 From Out of Old Fields Comes All This New Corn: An Historical Perspective on Heterosis in Plant Improvement

I. L. Goldman

University of Wisconsin
Madison, Wisconsin

ABSTRACT

Farming has always depended upon human-induced modifications of the natural world and heterosis, meaning *to alter* in Greek, is a prime example of technology in the service of agriculture. Unlike many key biotechnologies, however, its biological underpinnings remain poorly understood. Heterosis, a manifestation of the superiority of F1 performance relative to parental performance, is fundamentally concerned with inbreeding and outbreeding. The significance of outbreeding in the development of superior animals such as the mule has been noted since Mosaic times. Religious and moral teachings regarding human inbreeding likely preceded observations drawn from the experience of animal domestication by pastoral nations. Although relatively close inbreeding took place in early Egyptian and Greek cultures and among ruling classes in European societies, recognition of an incest taboo was commonplace in many of the world's cultures. By the 18th century, hybridizers were careful to note the paradox between the dangers of close inbreeding and wide outbreeding. Koelreuter was the first hybridizer to comment upon heterosis and the first to suggest it resulted from outbreeding. Knight, Mendel, and Darwin made significant contributions to the understanding of these phenomena during the 19th century, however they did not recognize the beneficial effects of outbreeding as opposite to the detrimental effects of inbreeding. Beal was influenced by Darwin and among the first to report increased yields with hybridization. Beal also trained Davenport and Holden who, along with East at the University of Illinois in 1900 pioneered inbreeding theory. Prior to this period, the primary aim of breeding programs was avoidance of inbreeding; thus hybrid vigor was said to work via the prevention of self-pollination. Breeders in this period focused on mass selection as inspired by Darwin's *selection principle*. Inspiration for the study of inbreeding may have come indirectly from an association between East and the botanist Hottes, recently back from study in Europe with Correns. The rediscovery of Mendel's laws by Correns and others may have influenced East to think of line purification from inbreeding. It was not until a landmark article by Shull in 1908 that the fundamental principle of inbreeding and its role in heterosis was clarified. In this work, Shull argued persuasively that selfing isolated homozygous lines and that these lines could be crossed to capitalize on heterosis; a fact that influenced East and helped to overturn the

---

Copyright © 1998 Crop Science Society of America, 677 S. Segoe Rd., Madison, WI 53711, USA.
Concepts and Breeding of Heterosis in Crop Plants. CSSA Special Publication no. 25.

negative image of inbreeding to a positive benefit obtained by hybridizing inbreds. Shull coined the word heterosis in 1914. Commercially feasible F_1 corn hybrids were developed following Jones' (East's student) 1918 proposition of the double cross. Since the 1930s, exploitation of heterosis has spread to include many other crop plants. For many crops, the full benefit of heterosis has yet to be realized because inbreeding efforts are still in their infancy from a genetic point of view. Although some have criticized the move toward F_1 hybrids as driven more by profit rather than science, this argument does not address other potential benefits of F1 hybrids such as uniformity. Use of heterosis has fostered the development of a worldwide seed industry and dramatically altered the landscape of professional plant breeding, undoubtedly making a substantial contribution to research and development efforts in this discipline. The success of F_1 hybrid technology has affected increases in food production in many regions of the world.

HISTORY OF INBREEDING AND OUTBREEDING

Historical treatments on heterosis (East, 1908; East & Jones, 1919; Jones, 1918; Zirkle, 1952; Shull, 1952; Hayes, 1952; Stuber, 1994) have reviewed in detail the creation of a modern heterosis concept. Much of the discussion presented herein is therefore concerned principally with those who recognized and exploited the phenomenon to the betterment of agriculture, and in particular those who contributed to our knowledge of inbreeding and outbreeding.

Substantial evidence exists to support the notion that early agriculturists were aware of the significance of inbreeding and outbreeding in both plant and animal species (Clutton-Brock, 1992). Perhaps the first significant example in this regard is the development of the mule. A mule results from a cross between the donkey (*Equus asinus*) and the horse (*Equus caballus*). Mules exhibit significant heterosis for size, strength, and endurance; all of which were integral to the development of modern animal agriculture. The Sumerians were producing these crosses at least 3000 years before the common era, indicating that hybrid technology may be at least 5000 years old (Clutton-Brock, 1992). In fact, the Sumerians and others in the Middle East and Mediterranean regions were producing donkey × onager (*Equuus hemionus*) crosses and making donkey × zebra mules (*E. asinus* × *E. caballus*) for the tropics; thus demonstrating that an animal breeding program to exploit heterosis was in place many thousands of years ago (Clutton-Brock, 1992).

Anthropological and genetic evidence also suggest that native Americans may have practiced hybrid technology. Although the cultivation of corn (*Zea mays* L.) by native American peoples has been well documented, little scientific information on their methods of selection and breeding are known. A remarkable article written in 1909, by G.N. Collins, then Assistant Botanist at the U.S. Department of Agriculture's Bureau of Plant Industry, suggested that these early Americans purposefully exploited hybrid technology:

> Among a number of primitive tribes where the cultivation of corn has reached a high state of development, the injurious effect of this close breeding appears to have been recognized, since they have methods of guarding against it. Thus the Indians in the region of Quezaltenango, in western Guatemala, and the Hopi Indians of Arizona make a regular practice of placing seeds of more than one local variety in each hill, with the idea that larger yields can be obtained in this way (Collins, 1909).

The close planting of diverse strains clearly indicates that a recognition of the positive benefits of heterosis was firmly in place many thousands of years ago in the Americas. In addition to this anecdotal comment, some research has been conducted to assess whether ancient corn from these regions fits a genetic pattern of modern-day corn hybrids. Helentjaris (1988) described the results of experiments with 700-yr-old corn remains obtained from several Anasazi sites in the southwestern USA. The identical restriction fragment length polymorphism (RFLP) patterns for each sample across sites and multiple restriction fragments detected in this study were suggestive of the kind of hybrid uniformity and genetic heterozygosity present in the RFLP patterns of modern-day corn hybrids. Although these data do not prove the purposeful construction of corn hybrids by the Anasazi, they do raise the interesting issue that technology designed to exploit heterosis may have been in place for many years before the modern scientific era would indicate. Archeologists and anthropologists have suggested that the Anasazi probably isolated certain corn populations or varieties for spiritual reasons (Helentjaris, 1988); thus it is feasible that isolation practices may have led to or been formulated out of observations of heterosis in food crops such as corn.

Ideas about inbreeding and outbreeding have evolved dramatically during human evolution (Frazer, 1941). Investigation of their foundations in human culture provides a glimpse into the power of these two reproductive modes and shows how they were closely connected to the development of religion and society. In the main, ideas about totemism, nature worship, and other forms of idolatry were designed to connect human reproductive behavior with those behaviors observed in nature (Frazer, 1941). This in turn led to a variety of reproductive rites imposed by human populations on themselves and on the natural populations upon which they depended. Expressions of these rites may have included the bringing together of diverse plants and animals for mating rituals and the separation of various plant and animal populations for purification reasons. Epic works such as the Golden Bough described how early human societies modeled their reproductive behaviors after those observed in nature or those sought in agriculture for the purposes of good harvests and other forms of positive spiritual and material consequences. Finally, Sigmund Freud and others recognized and wrote about the widespread acceptance of an incest taboo among many of the world's cultures, indicating that humans have known for many thousands of years about the consequences of inbreeding and outbreeding (Freud, 1912).

EARLY PLANT HYBRIDIZERS

Plant hybridizers in the 18th century were among the first to make significant scientific contributions to the development of a modern heterosis concept. Much of their work was concerned more with experimentation with the results of divergent crosses rather than practical plant improvement (Mayr, 1982). Cotton Mather was the first to note and describe the xenia effect, or the immediate effect of the pollen parent on the female seed parent (Mayr, 1982). This casual observation made a significant contribution to pollination biology and was an important conceptual advance in plant hybridization. In 1776, Kolreuter became the first of the plant hybridizers to document in detail the results of his crosses, in this case in the genus *Nicotiana*, and to describe significant heterosis (Mayr, 1982). Knight took these ideas one step further and, in 1799, suggested that the widespread existence of cross-pollination in nature was proof that nature *intended* this to be the norm (Mayr, 1982). In 1828, Wiegmann described the results of crosses leading to heterosis in the crucifers. Gartner and Focke, in 1849 and 1881, respectively, detailed the results of their crosses, noting heterosis, and encouraging other scientists to think along the lines of enhanced plant growth through such hybridization (Mayr, 1982). Mendel (1865), described not only the fundamental laws of heredity thus founding the science of genetics, but also detailed the results of his crosses in doing so. Mendel commented on the luxurious growth observed in hybrids and, because of the eventual influential nature of his work, contributed additional observations to the development of a heterosis concept.

We now turn to the debt we owe Charles Darwin, the contributor of so many important ideas in biology, for his pioneering work in the area of plant hybridization. Darwin had completed a typically painstaking analysis of reproductive modes in plants and concluded:

> Nature thus tells us, in the most emphatic manner, that she abhors perpetual self-fertilization (Darwin, 1862).

Darwin detailed the results of 37 crosses including corn in which he observed increased height in 24 (Darwin, 1877). He commented on the decrease in height observed in the self-pollinated plants and discussed many natural mechanisms by which plants avoid inbreeding. These observations became critical in developing an understanding of the significance of inbreeding and outbreeding in nature. Interestingly, Darwin noticed that the deleterious effects of inbreeding could be *reversed* following the crossing of inbred strains. In the cases he tested, the performance of the first cross-pollinated generation was quite vigorous. As such, he suggested that the effects of inbreeding were essentially reversible upon intermating. This hints quite closely at a modern concept of the relationship between inbreeding and outbreeding with respect to heterosis, however Darwin's ideas did not draw a clear path between the effects of inbreeding and those of outbreeding.

EARLY INBREEDING AND OUTBREEDING THEORY

Perhaps Darwin's other major contribution in this area was to serve as an inspiration to Harvard botanist Asa Gray. Gray was an important figure in the biological sciences at the time in the USA and he was Darwin's correspondent. The two had visited in England several times and shared ideas on inbreeding and outbreeding, and this correspondence may have had a very significant impact on the paths taken by Gray in his research. Clearly, correspondence between Darwin and Gray established Darwin's prior claim on the details of natural selection prior to the joint publication of Darwin and Wallace in 1859 (Wallace & Brown, 1988; Darwin & Wallace, 1859). Gray in turn served as the research mentor for William James Beal (Fig. 1–1), who earned his degree at Harvard in 1865 and became one of the first practical contributors to the modern heterosis concept (Wallace & Brown, 1988; Crabb, 1947). Beal's career at Michigan Agricultural College (later Michigan State University) focused on varietal crossing and pollination control in corn; two practices with great influence on early plant hybridizers (Wallace & Brown, 1988). Beal then served as the inspiration and mentor of Perry Holden and Eugene Davenport, who made great strides in early inbreeding theory at the University of Illinois at Urbana-Champaign (Wallace & Brown, 1988). This latter group also contained chemist Cyril Hopkins, who began the Illinois Long Term Selection experiment in 1896, and later E.M. East, who was to become perhaps the most important and influential figure in development of modern scientific plant breeding in the USA.

Beal's contribution to the development of a heterosis concept was an emphasis on pollination control as a means of corn breeding Champaign (Wallace & Brown, 1988). The idea here was to minimize self-pollination, for its deleterious effects had been carefully noted by many workers beginning with Darwin. Early corn breeding methods, heavily influenced by the work of Beal, thus focused on ways to prevent inbreeding through pollination control rather than exploiting specific matings in the development of cross-bred strains. During this period, Beal spearheaded a widespread effort to examine the deleterious effects of inbreeding. Archibald Shamel and Eugene Funk both began inbreeding pro-

Fig. 1–1. Academic-influence pedigree chart from Charles Darwin to D.F. Jones showing a flow of ideas about inbreeding and outbreeding that led to a modern heterosis concept. Specific contributions of certain individuals are noted in smaller type under their names. The central column represents the flow of ideas from worker to worker. The left column represents the country or state where the ideas were formulated and the right column represents the year or general era in which the ideas were first discussed.

grams at the University of Illinois, only to stop them in the early part of the century due to the widespread feeling enough damage had been done by inbreeding already. Davenport's words at the time may have summed up the sentiment on inbreeding:

> The effects of inbreeding appear both pronounced and disastrous; the second generation from inbred seed being less than two-thirds normal size and nearly barren . . . but the second planting from this seed when closely selected after the same plan left almost a full stand, which shows that corn may be brought much nearer a constant type than has ever yet been done (Davenport, as cited in Fitzgerald, 1990).

A CULTURE OF INBREEDING: CORN BREEDING AT THE UNIVERSITY OF ILLINOIS

Following the passage of the Hatch Act establishing agricultural experiment stations at land-grant universities in 1887, much practical corn breeding work began to take place at these institutions. Beal was active in training students in corn breeding. Both Eugene Davenport (M.S. degree, 1884) and Perry Holden (M.S. degree, 1895) learned pollination control theory and techniques from Beal at Michigan State (Wallace & Brown, 1988). The work conducted by Beal and his students was repeated and confirmed at the University of Illinois Agricultural Experiment Station by George Morrow and F.D. Gardner. This work demonstrated the positive benefits of hybridization in terms of improved yield (Morrow & Gardner, 1893; Morrow & Gardner, 1894). These workers not only emphasized the importance of cross pollination but recommended the alternate planting of varieties with detasseling in order to produce hybrid seed. A section in their 1894 manuscript entitled 'Results from Cross-Bred Corn' (Morrow & Gardner, 1894) is perhaps the first outline of the hybrid corn breeding method still in use today (Troyer, 1996, personal communication). Eugene Davenport was appointed Dean of the College of Agriculture at the University of Illinois and he later appointed Perry Holden as a Professor in the College of Agriculture in 1896. In the same year, Cyril Hopkins initiated the Illinois Long-Term Selection experiment, the longest-running crop selection experiment in modern plant breeding history. E.M. East was hired by Hopkins to assist with the Long-Term Selection project. East had been a student in a botany course with Charles Hottes, who had just returned from European study with the rediscoverers of Mendel's laws. Hottes was therefore attuned to the hereditary foundation by which inbred lines might be developed from segregating populations. It has been suggested that East may have been influenced to think about line uniformity and other such aspects of the genetic foundations of inbreeding through his contact with Hottes. Hottes remarked:

> I liked East very much. He was a good student. He didn't have to be driven, although he was a rather retiring pleasant sort of fellow who as a youth gave rather little indication of the fine qualities of aggressiveness which he developed later. He always constructed his sentences with great care, speaking in a rather thin high voice. He was many times studious to the point of being preoccupied (Hottes, as cited in Fitzgerald, 1990).

The Illinois Long-Term Selection experiment was influenced primarily by work conducted in France by early breeders of the fodder beet (Hopkins, 1899). When the Napoleonic Wars left France without an inexpensive source of sugar, Napoleon offered a prize for a new European source of sugar (Troyer, 1996). The discovery by Andreas Marggraf in 1747 that fodder beets (*Beta* sp.) produced sucrose identical to that of sugar cane led to late 18th century mass selection efforts by Franz Karl Achard to increase sugar concentration in this crop (Duvick, 1996). This work was followed by the landmark pedigree breeding program of Louis de Vilmorin, grandson of the founder of the Vilmorin Seed Company. Vilmorin's work focused on progeny testing and thus measured the breeding potential of parental lines by assessing the performance of their crossed progeny. This work necessarily involves accurate pedigree records, and thus both pedigree breeding and progeny testing evolved in 19th century France, supplanting the previous mass selection efforts, in the context of improving a quantitative trait (sucrose content) in beets. Achard and Vilmorin were important influences on Hopkins (Hopkins, 1899) and he based his modification of chemical composition of the corn kernel on these principles (Fig. 1–2). The change in emphasis at the University of Illinois away from inbreeding research and on to mass selection and progeny testing may have delayed the development of a modern heterosis concept. Hopkins said to East:

> We know what inbreeding does and I do not propose to spend people's money to learn how to reduce corn yields (Hopkins, as cited in Fitzgerald, 1990).

Fig. 1–2. Academic-influence pedigree chart from Buffon to Shull showing a flow of ideas about inbreeding and outbreeding that led to a modern heterosis concept. The central column represents the flow of ideas from worker to worker. Specific contributions of certain individuals are noted in smaller type under their names. The left column represents the country or state where the ideas were formulated and the right column represents the year or general era in which the ideas were first discussed.

E.M. East, who went on to uncover some of the key aspects of the modern heterosis concept, was clearly interested in pursuing inbreeding research at Illinois. It is possible that the shift in emphasis away from this work at Illinois led to East's departure for the Connecticut Agricultural Experiment Station and may have delayed some of his early inbreeding work. Although this work was conducted while East was at Connecticut, precedence for the formulation of the modern heterosis concept is often given to G.H. Shull.

COMPOSITION OF A FIELD OF CORN

Turning to the final academic-influence pedigree (Fig. 1–3), I introduce the key progression of thoughts that led to the development of a modern heterosis concept. Both Buffon and Lamarck in France offered key insights into the workings of evolution and natural selection. These ideas influenced early plant hybridizers such as de Candolle and Nageli, both of whom made important contributions to early ideas about the results of hybridization (Mayr, 1982). These workers in turn served as the inspiration for the rediscoverers of Mendel: Correns, De Vries, and Von Tschermak. In particular, the important mutation theories developed by De Vries were of interest to the young George Harrison Shull, a brilliant young biologist who earned his graduate degree at the University of Chicago. Shull wished to test some of De Vries' ideas about the generation of

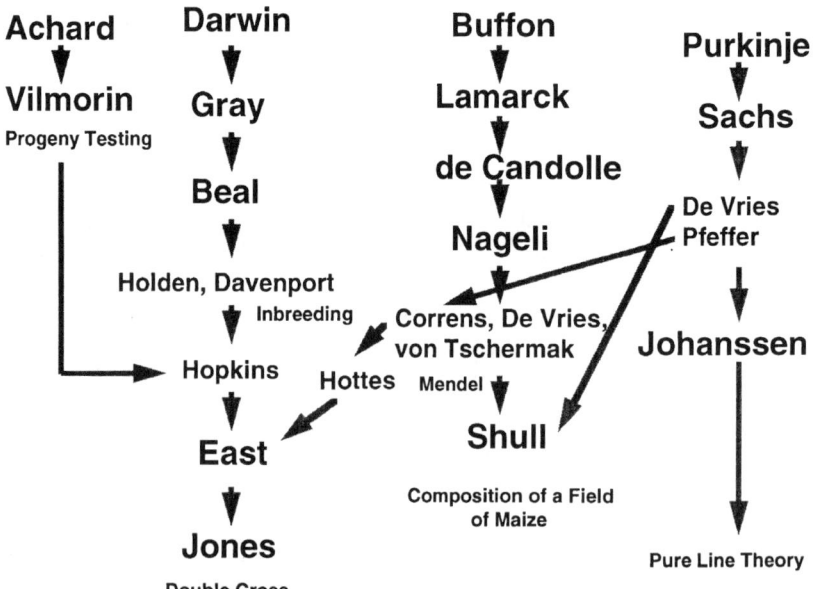

Fig. 1–3. Combined academic-influence pedigree chart showing influences of four streams of thought on the development of a modern heterosis concept. Specific contributions of certain individuals are noted in smaller type under their names.

variation by mutation, however his interest was primarily directed at basic genetic mechanisms and not crop improvement. Interestingly, De Vries and Pfeffer, both influenced by Sachs and Purkinje (Fig. 1–3), influenced Johannsen, who made a significant contribution to inbreeding theory and quantitative inheritance by developing the pure-line theory (Johannsen, 1903; Sturtevant, 1965). Upon assuming his first position in 1904 at Cold Spring Harbor Laboratories in Long Island, NY, Shull began inbreeding a number of crop plants, including corn. The next 4 yr were consumed with inbreeding and outbreeding work in a variety of species. Ultimately, Shull developed a perspective on heterosis that he outlined in a 1908 publication entitled 'Composition of a Field of Maize.' This article is considered to be the landmark development in early heterotic theory, as it clearly states that a corn variety is a complex mixture of genotypes. Shull commented that since each plant is, in a sense, isolated by inbreeding; each plant is of an essentially different genotype. He observed and measured a reduction in vigor due to the segregation of these different types into their respective homozygous classes and showed how the F_1 yield from crosses of these types exceeded the parental varieties from which they came (Shull, 1908). The significance of this landmark article is that Shull was able to bring together the key aspects of inbreeding and outbreeding theory and show how they are related in a coherent heterosis concept (Shull, 1952).

E.M. East and G.H. Shull are often given credit for developing key aspects of the modern heterosis concept. Although both men clearly contributed crucial ideas to this field, Shull is thought of as having claim to the first publication of the idea, based on the following evidence. East had discussed the dangers of inbreeding as late as the 1907 Connecticut Agricultural Experiment Station Bulletin (East, 1908); however, he was heavily influenced by Shull's presentation at the American Breeder's Association Meeting in Chicago in early 1908. In the 1908 Connecticut Agricultural Experiment Station Bulletin, East stated:

> A recent paper by Dr. G.H. Shull has given, I believe, the correct interpretation of this vexed question. His idea, although clearly and reasonably developed, was supported by few data (East, 1908).

Finally, in early 1908, East wrote to Shull and said:

> Since studying your paper, I agree entirely with your conclusion, and wonder why I have been so stupid not to see the fact myself (Shull, 1952).

East is also to be given a great deal of credit for his extensive training of leaders in the field of plant breeding. Many of the students trained by East at the Bussey Institute at Harvard went on to become influential scientists and plant breeders in the USA and abroad. For example, East trained (i) D.F. Jones, popularizer of hybrid corn by suggesting the double cross in 1918; (ii) R.A. Emerson, mentor for many of plant breeding's early pioneers at Cornell University; (iii) E.R. Sears, pioneer of chromosome manipulations in plant breeding at Missouri; (iv) C.M. Rick, developer of one of the world's most remarkable collections of crop genetic material at the University of California, Davis; (v) R.A. Brink and H.K. Hayes, pioneer teachers and researchers at the University of Wisconsin and University of Minnesota, respectively; and (vi) P.C. Mangelsdorf of Texas A&M University and Harvard, crop evolutionist (Troyer, 1996).

ACCIDENTS AND AGENDAS IN MODERN HETEROSIS

Some have questioned whether the development of a modern heterosis concept would have come about so rapidly without corn as a model organism. Indeed, corn is thought to be central to the development of heterotic theory because it lent itself so well to experiments involving inbreeding and outbreeding. Simmonds (1979) suggested three reasons why corn was so suitable: (i) The development of hybrid corn came along at a time when new methods of corn breeding were needed. In fact, the early workers at Illinois were charged with the development of new corn breeding methods, and thus there was likely support in both the scientific and agricultural communities for such investigation. (ii) The development of agriculture in the USA was closely tied to the development of corn as a staple crop. The sheer economic weight of corn during the early part of the 20th century was responsible for fueling the development of new and productive varieties. (iii) Finally, and perhaps most important, it has been suggested that the relatively simple emasculation techniques practiced in corn breeding and the simple monoecious reproductive structure of the corn plant allowed for easy inbreeding and outbreeding. This in no small way helped to make corn the model organism for the study of these two key phenomena. In addition to these reasons, it is evident that the effects of inbreeding and outbreeding were more dramatic in corn than in other grain crops such as wheat (*Triticum aestivum* L.), oat (*Avena sativa* L.), or rice (*Oryza sativa* L.). Thus, this organism provided perhaps the best opportunity for study of the effects of these two breeding methods.

In addition to improved yield performance, many other benefits have been derived from hybrids and the exploitation of heterosis. One of the most often overlooked benefits is uniformity, an element that has certainly allowed for rapid expansion of production in many crop plants such as the vegetables. This characteristic of hybrids has enabled the production of more uniform vegetable crops for fresh market production, contributing to enhanced consumer appeal and greater market value. Furthermore, the benefits of uniformity with respect to maturity have been associated with greater efficiencies during harvest. The development of mechanical harvesting technology for major grain crops such as corn in the 1920s coincided perfectly with the development and widespread use of hybrid corn. These two events are connected in that the added benefit of uniformity obtained from hybrids allowed for widespread dissemination of mechanical harvesting technologies. Additional benefits may include stress tolerance and pest resistance and other performance characteristics. Severe droughts in the U.S. Cornbelt during 1934 and 1936 resulted in poor corn crops; however, hybrids often out-performed open-pollinated cultivars under these conditions.

In his landmark 1908 article, Shull discussed the *nonrenewability* of hybrids, and East and Jones (1919) had commented that the advancement of hybrid technology would provide less incentive for individual breeders to improve open-pollinated cultivars. The dramatic shift toward hybrid technology revolutionized many sectors of the agricultural economy. One sector that was affected greatly was the seed industry, because the adoption of hybrids meant being able to sell seed to farmers year after year. Berlan and Lewontin (1986) have criticized the move toward hybrids as merely an expression of class interest. They further suggest that improvement of open-pollinated cultivars through careful selection should have resulted in agronomic performance on a par with hybrid varieties. Kloppenburg (1989) has even called hybrid corn agriculture's Manhattan project. The intriguing perspective that genetic gain in open-pollinated cultivars might have rivaled those of hybrids has, in some cases, been supported experimentally (Duvick, 1977, 1992, 1996; Hallauer & Miranda, 1981). Despite this contention, others have suggested that many of these arguments against hybrid technology reveal their political agenda by expressing surprise that clever individuals learned to profit from agriculture (Goodman, 1989). One of the major agricultural changes to emerge from hybrid technology, therefore, has been a flow of germplasm and control of germplasm from the public to the private sector; a situation that has aroused a great deal of interest in both the scientific community and the general population.

As the academic great-great grandchildren of E.M. East and George Shull make their contributions to the science of plant breeding at the dawn of the 21st century, they will hopefully do so out of respect for the foresight and determination of those who paved the way. It is perhaps the goal of the historian to renew our appreciation of the past, and so in this tradition I offer you, in the words of Chaucer: out of these old fields, all this new corn.

ACKNOWLEDGMENTS

I am indebted to my colleague, William F. Tracy, for many helpful discussions and for bringing to my attention the article by Collins. I also am grateful to Forest Troyer, Cargill Hybrid Seeds, for many important contributions on the history of corn breeding at the University of Illinois and for helpful discussions during the preparation of this manuscript.

REFERENCES

Berlan, J.P, and R. Lewontin. 1986. The political economy of hybrid corn. Monthly Rev. 38:35–47.

Clutton-Brock, J. 1992. Horse power. Harvard Univ. Press, Cambridge, MA.

Collins, G.N. 1909. The importance of broad breeding in corn. USDA Bureau of Plant Industry Bull. 141. Part 4:33–42. U.S. Gov. Print. Office, Washington, DC.

Crabb, A.R. 1947. The hybrid-corn makers: Prophets of plenty. Rutgers Univ. Press, New Brunswick, NJ.

Darwin, C. 1862. On the various contrivances by which British and foreign orchids are fertilised by insects, and on the good effects of intercrossing. Murray, London.

Darwin, C. 1877. The effects of cross and self-fertilization in the vegetable kingdom. Appleton, New York.

Darwin, C., and A. Wallace. 1859. On the tendency of species to form varieties; and on the perpetuation of varieties and species by natural means of selection. J. Linn. Soc. London. (Zool). 3:45–62.

Duvick, D. 1977. Genetic rates of gain in hybrid maize yields during the past 40 years. Maydica 22:187–196.

Duvick, D. 1992. Genetic contributions to advances in yield of U.S. maize. Maydica 37:69–79.

Duvick, D. 1996. Plant breeding, an evolutionary concept. Crop Sci. 36:539–548.

East, E.M. 1908. The relation of certain biological principles to plant breeding. Conn. Agric. Exp. Stn. Bull. 158. State of Connecticut, Hartford.

East, E.M., and D.F. Jones. 1919. Inbreeding and outbreeding. J.B. Lippincott Co., Philadelphia.

Fitzgerald, D. 1990. The business of breeding: Hybrid corn in Illinois, 1890–1940. Cornell Univ. Press, Ithaca, NY.

Frazer, J.G. 1941. The golden bough. Macmillan, New York.

Freud, S. 1912. Totem and taboo. *In* P. Gay (ed.) The Freud reader. 1989. W.W. Norton & Co., New York.

Goodman, M.M. 1989. Diversity 5(1):33–35.

Hallauer, A.R., and J.B. Miranda. 1981. Quantitative genetics in maize breeding. Iowa State Univ. Press, Ames.

Hayes, H.K. 1952. Development of the heterosis concept. p. 49–65. *In* J.W. Gowen (ed.) Heterosis. Iowa State College Press, Ames.

Helentjaris, T. 1988. Does RFLP analysis of ancient Anasazi samples suggest that they utilized hybrid maize? Maize Newsl. 62:104–105.

Hopkins, C.G. 1899. Improvement in the chemical composition of the corn kernel. Illinois Agric. Exp. Stn. Bull. 55:205–240.

Johannsen, W. 1903. Ueber Erblichkeit in Populationen und in reinen Linien. Gustav Fischer, Jena. Translated version (Heredity in populations and pure lines) appears in Classic papers in genetics. 1959. J.A. Peters (ed.) Prentice-Hall, Englewood Cliffs, NJ.

Jones, D.F. 1918. The effects of inbreeding and crossbreeding upon development. Connecticut Agric. Exp. Stn. Bull. 207. State of Connecticut, Hartford.

Kloppenburg, J.R., Jr. 1988. First the seed. Cambridge Univ. Press, New York.

Mayr, E. 1982. The growth of biological thought. Belknap Press, Cambridge, MA.

Morrow, G.E., and F.D. Gardner. 1893. Field experiments with corn, 1892. Univ. of Illinois Agric. Exp. Stn. Bull. 25:173–203.

Morrow, G.E., and F.D. Gardner. 1894. Field experiments with corn, 1893. Univ. of Illinois Agric. Exp. Stn. Bull. 31:333–360.

Shull, G.F. 1908. The composition of a field of maize. Rep. Am. Breed. Assoc. 5:51–9.

Shull, G.F. 1952. Beginnings of the heterosis concept. p. 14–48. *In* J.W. Gowen (ed.) Heterosis. Iowa State College Press, Ames.

Simmonds, N.W. 1979. Principles of crop improvement. Longman, London.

Stuber, C.W. 1994. Heterosis in plant breeding. p. 227–251. *In* J. Janick (ed.) Plant breeding reviews. John Wiley & Sons, New York.

Sturtevant, A.H. 1965. A history of genetics. Harper & Row, New York.

Troyer, F. 1996. Early Illini corn breeders: Their quest for quality and quantity. p. 56–67. *In* Am. Seed Trade Assoc. Hybrid Corn–Sorghum Res. Conf., Chicago, IL. 5–6 Dec. 1995. Dolores Wilkinson, Washington, DC.

Wallace, H.A., and W.L. Brown. 1988. Corn and its early fathers. Revised ed. Iowa State Univ. Press, Ames.

Zirkle, C. 1952. Early ideas on inbreeding and crossbreeding. p. 1–13. *In* J.W. Gowen (ed.) Heterosis. Iowa State College Press, Ames.

2 Heterosis: Performance Stability, Adaptability to Changing Technology, and the Foundation of Agriculture as a Business

Dwight T. Tomes
*Pioneer Hi-Bred International
Johnston, Iowa*

ABSTRACT

The agriculture of the past century has been romanticized as an ideal of a simple, virtuous, and good life, and is a cornerstone of deeply held beliefs in America. Considered as a business, this agriculture featured small farms, labor intensive inputs (both human and animal) and minimal mechanical, chemical, and capital inputs. Hybrids were uniquely positioned to take advantage of many of the changes that occurred in the business of agriculture in this century: increased chemical inputs, mechanization, capitalization, farm size and decreased use of human and animal labor. Heterosis allowed more effective use of the available genetic resources in open-pollinated varieties. Interest in heterosis, specifically maize hybrids, also laid the foundation for private and public research, which fostered the continued development of improved maize hybrids. Yield stability, in addition to increased yield potential, were important stimulants to the capital investment in farming. During the normal years in the 1920s and 1930s when open-pollinated varieties predominated, about 85% of the maize (*Zea mays* L.) planted was harvested. During the drought stress years of 1934 and 1936 harvest rates dipped to only 61 and 67%. After widespread adoption of hybrids (1940 and beyond), the proportion of harvested hectares has fluctuated from 85 to 92%, regardless of the environment of any particular year; however, the yield in each of these high stress years *was* lower than that recorded in more normal years. Modern hybrid maize differs primarily from open-pollinated and earlier hybrids in their response to stress. Newer hybrids have improved water stress performance, are much less prone to silk delay, have significantly lower respiration rates during silking, have longer periods of grain fill, and are higher yielding under both low and high input environments. Tassel size has decreased as has premature death (stay green scores are improved) and stalk rot has decreased. Resistance to European Corn Borer 2nd brood has increased. Hybrids in other species, such as tomatoes (*Lycopersicon esculentum* L.), demonstrate other attributes that may have been of equal importance to adoption in agriculture. The two parents of a hybrid may combine unique combinations of traits not available in either single parent. Increased shelf life in fresh market tomatoes is one such example. The longer shelf life tomato may be analogous to yield stability in maize since harvest can occur over a longer time window within a given environment. Yield stability and better predictability have profound impli-

---

Copyright © 1998 Crop Science Society of America, 677 S. Segoe Rd., Madison, WI 53711, USA.
Concepts and Breeding of Heterosis in Crop Plants. CSSA Special Publication no. 25.

cations: Geographic areas that favor higher yields and returns will displace those areas in which yields are lower (total hectares for maize in the USA are lower today than in 1920). Business unit size (i.e., farm size) will increase as long as profitability within a commodity is higher. Choice of hybrids by farmers dictate both the scale of research and by whom research continues in hybrids, i.e., companies with larger market share tend to have more research.

"THE GOOD OLD DAYS"

The agriculture industry early in this century consisted of small farms in which human and animal labor was the major input, with minimal mechanization. Large families were a prominent feature of such farms because of the *home grown* source of labor. Maize was grown for livestock feed with relatively small amounts sold off farm. Anthia Huff of Nash, KY, describes the extended family nature of labor and some of the tasks in a letter to her husband away in the Army in WWI:

> " Tuesday night Sept. 10, 1918.: My Dear Husband: Your dear letter recd to-day, and was so glad to hear from you. I am feeling all right to-night, excepting being tired. I have been helping Cleve & Tina top corn to-day. We put up 48 big Shocks, and tied 75 bundles of fodder, but it was almost Sundown when we quit . . . . Thursday, Oct. 24th: My brother, Oscar, came down yesterday. He brought logs for the sawmill but the sawmill was down. He helped me gather some corn. We gathered down to where the corn was topped . . ."

The adoption of hybrid maize, begun in the late 1930s, was a new technology precisely timed to exploit changes that were occurring in American agriculture. Hybrids revolutionized the way maize was grown, and changed the expectations that farmers had for the crop (Ball & Heady, 1972b). The nature of the adoption of hybrids is best told by one who experienced hybrids first hand. A farmer, Thomas Embry, of Leitchfield, KY, recalled (June, 1996) his first look at hybrid maize:

> "In the 30s we raised open-pollinated corn, Tennessee Red Cob, one of the best ones available. Our county agent, Mr. R.T. Faulkner, wanted us to test a hybrid, Ky13 (perhaps US13). The University was doing experiments with hybrid seed corn. They also had a white hybrid, #72 white. We had a 22 acre field near our house. We tested a 150 foot section, 10 rows of our corn and 10 rows of the Ky13 and 10 rows of the 72 white. This test was repeated in four different sections. We normally planted corn in 22 inch spacing, and thinned to 20 inches because if the corn was planted thicker, the ears were small or the plants went barren. The county agent planted the hybrids at about a 10 inch or maybe a little closer spacing and asked us not to thin the two hybrids. When we gathered it that fall and weighed it, there was a big difference. We never did grow the open-pollinated corn any more after that, we always went to the hybrids. We planted the KY13 for many years, and then on to company hybrids . . ."

The heterosis observed in hybrid maize made use of the genetic resources of that time (open-pollinated varieties and land races) in a more productive manner. Specific genotypes (inbreds) from the open-pollinated source populations gave rise to hybrids of much higher yield than the open-pollinated germplasm.

INDEX OF FERTILIZER USE, FARM LABOR AND MECHANIZATION 1910-1968

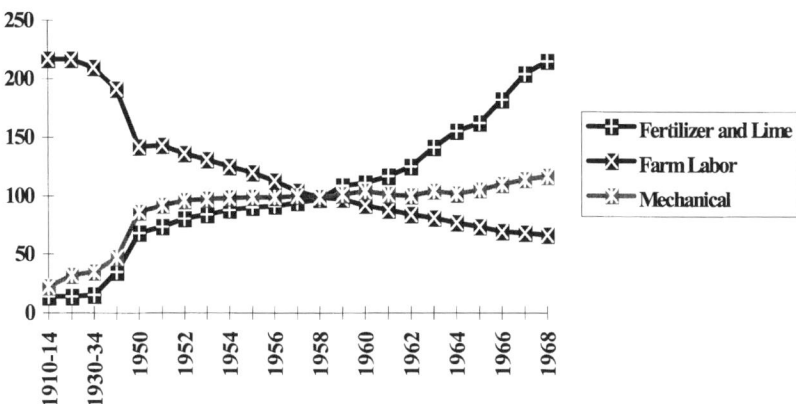

Fig. 2–1. Index of fertilizer use, farm labor, and mechanization from 1910–1968. Source: U.S. Department of Agriculture (1969).

Industrialization, accelerated by World War II, impacted agriculture in a dramatic fashion: (i) labor was drawn from agriculture to better paying jobs in urban areas; (ii) inputs such as N fertilizer became widely available; (iii) mechanization began to replace human and animal labor; (iv) the supply of land for agriculture became a finite resource; and (v) the transportation infrastructure had developed to support agricultural marketing of crops in a more effective manner (USDA, 1969; Fig. 2–1).

YIELD STABILITY

Farmers observed (see above) that hybrid maize could be grown at a higher population, and under stress conditions where open-pollinated maize would have been barren. The increasing yield of hybrids over the years has been very well documented by Duvick (1984, 1996b; see below) and others (Table 2–1); however, the increased consistency and predictability of hybrids has been as important as yield increases. Madden and Partentheimer (1972) have proposed that farm size in the early part of the century was probably limited by uncertainty. The two major contributors of this uncertainty were the lack of price support and susceptibility of open-pollinated maize to low yields because of stress. In the 1930s, government price support programs reduced price uncertainty and hybrid maize as well as other inputs reduced yield uncertainty.

Maize yields have risen dramatically during the past 50 yr, and by inference the consistency of yield also may have improved. Comparison of planted versus harvested hectares of maize over time is a straightforward way to view the consistency of harvestable yield. The proportion of harvested versus planted hectares

Table 2–1. Summary of characteristics of more recent corn hybrids compared with older hybrids and open pollinated varieties.

Trait	Comparison	New hybrid change	Reference
Barrenness	Era	Newer hybrids less barren	Duvick, 1984, 1992
Yield under weed competition	Two hybrid	Newer hybrids 21% higher yield	Tollenaar et al., 1997
Yield: low-high N	Two hybrid	Newer hybrids yield more	Tollenaar et al., 1997
Yield: low N	30–80s era	2310 kg ha^{-1}, mean yield increased 65%	Castleberry et al., 1984
Yield: high N	30–80s era	4690 kg ha^{-1}, mean yield increased 61%	Castleberry et al., 1984
Yield at different densities 10–79 kg ha^{-1}	Four hybrid, 40–80s era	Higher yield at higher density; increase 8–93 kg ha^{-1} yr^{-1} yield	Tollenaar et al., 1992; Duvick, 1992
Duration of grain fill	Era	Longer grain fill period (later)	Cavalieri & Smith, 1984
Water stress at silking	Two hybrid	Newer hybrid 25% lower stem water potential	Nissanka et al., 1996
Water stress transpiration silking	Two hybrid	Canopy transpiration newer hybrid 31% higher during recovery day	Nissanka et al., 1996
Canopy photosynthesis during water stress	Two hybrid	Newer hybrid 10–25% higher canopy photosynthesis	Nissanka et al., 1996
Respiration rate water stress	Two hybrid	Newer hybrid 30% lower respiration	Nissanka et al., 1996
Stay green	40–80s era	Newer hybrids live longer	Cavalieri & Smith, 1984; Duvick, 1984, 1992
Root lodging	4080 era	Newer hybrids improved	Duvick, 1984, 1992
Stalk lodging	40–80s era	Newer hybrids improved	Duvick, 1984, 1992
Upright leaf habit	40–80s era	Newer hybrids more upright leaf habit	Duvick, 1984, 1992
Tassel size, branches	40–80s era	Newer hybrids smaller tassel, fewer branches	Duvick, 1992
Grain composition: protein, starch	40–80s era	Newer hybrids less protein (10.2 to 9.0%) more starch	Duvick, 1992
European Corn Borer, 2nd brood	40–80s era	Newer hybrids better	Duvick, 1992
Kernel number and plant growth	Ontario era	More kernels/plant and higher growth rate during silking	Tollenaar et al., 1992
Anthesis-silking interval	40–80s era	Newer hybrids have stronger silk growth	Duvick, 1992
Tillering	40–80s era	Newer hybrids have less tillers	Duvick, 1992

increased during the period in which hybrid maize was widely adopted in mid century (Fig. 2–2a). The proportion of harvested land has increased from approximately 80% in the 1930s to about 90% since 1980. Most importantly, the fluctuation has been reduced dramatically. This reduction can be best visualized by comparing two 10-yr periods: 1927–1937 and 1985–1995 (Fig. 2–2b and 2–2c) in which both good and stress years can be compared. During most of the period from 1927–1937, about 85% of the planted hectares were harvested except for 1934 and 1936, which were drought years in which only 64 and 67% of planted maize was harvested. Comparison of 1985–1995 indicates that approximately 90% of the maize area was harvested despite extreme moisture deficit in 1988,

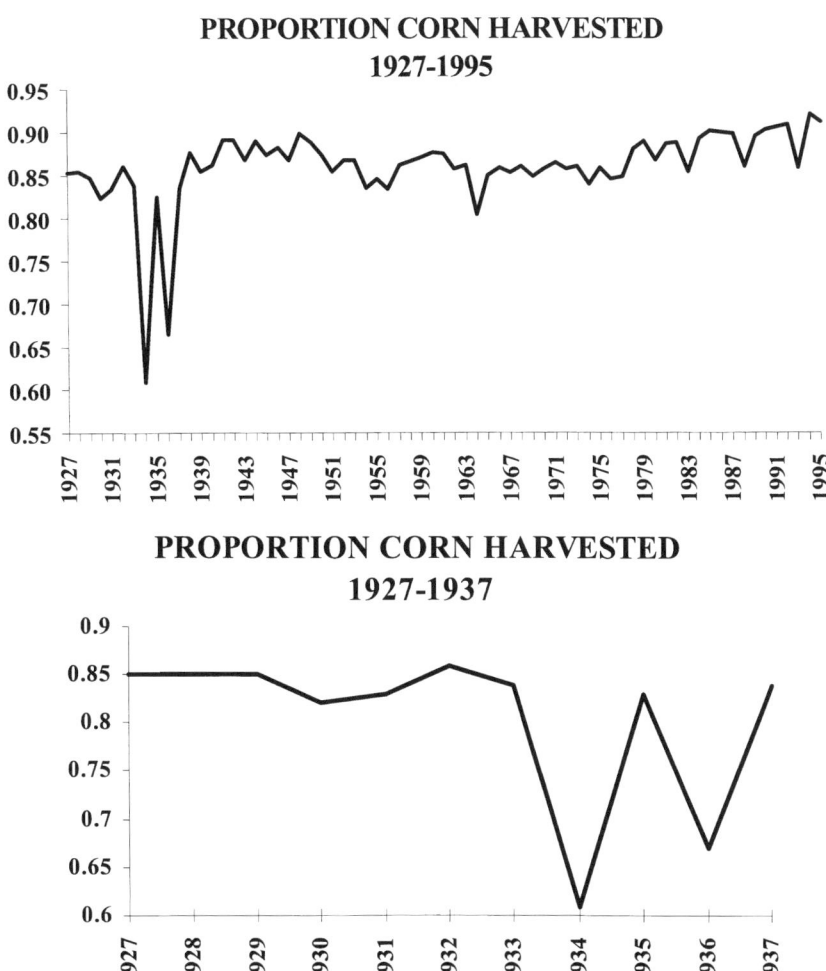

Fig. 2–2. Proportion of maize hectares harvested (a) 1927–1995; (b) 1927–1937; (c) 1985–1995. Source: U.S. Department of Agriculture/National Agricultural Statistics Service (1996). (Continued on next page.)

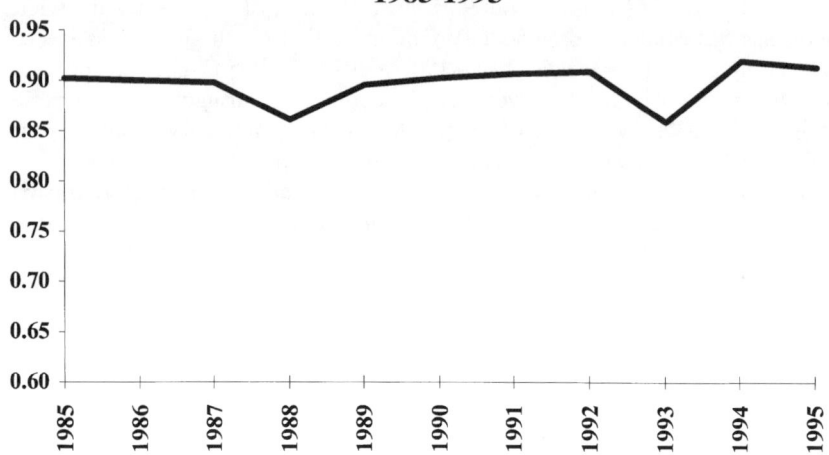

Fig. 2–2. Continued.

and excess moisture of 1993. Although yields were depressed in 1988 and 1993, most of the maize crop was harvested.

The number of hectares devoted to maize in the USA has declined from the 1920s (Fig. 2–3). The record hectares for maize in the USA was in 1917 when farmers harvested about 45 million ha. The record yield year was 1994 (9.4 T ha^{-1}) in which about 30 million ha were harvested (USDA-NASS, 1996). If yields were the same as the decade of 1927–1937, approximately 136 million ha would have been required to produce the maize harvested in 1994—almost five times the land devoted to maize in 1994!

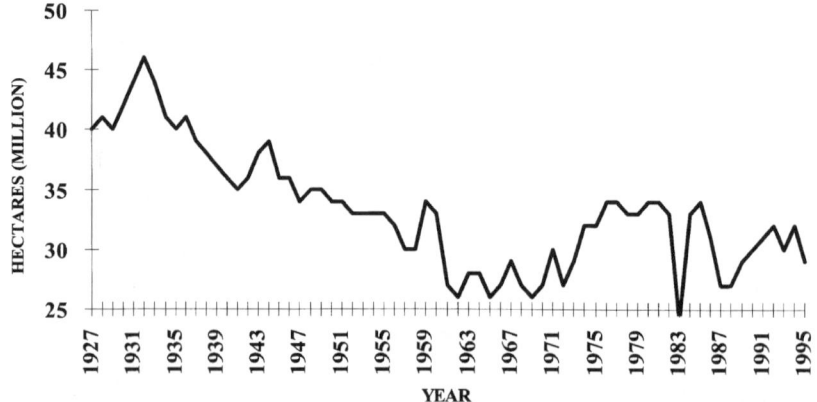

Fig. 2–3. USA grain maize hectares from 1927 to 1995. Source: National Corn Growers Association, 1995).

BREEDING HAS CHANGED THE MAIZE PLANT

Yields of todays hybrids are similar to the open-pollinated varieties grown in the 1920s and 1930s only when grown under very low stress conditions; however, the agronomic conditions necessary for productive agriculture throughout this century *have been and are stress filled!* Todays hybrids exceed the performance of open-pollinated varieties and earlier hybrids in many, if not all, comparisons (Table 2–1). Barrenness and premature death have been reduced and yield has increased under stresses such as high population density, weed competition, low and high water stress and with both low and high nitrogen levels (Table 2–1; Duvick 1984, Duvick 1996b; Castleberry et al., 1984; Tollenaar 1992; Cavalieri & Smith, 1984; Nissanka et al., 1996). Increased productivity is apparent under both intense management such as high N and low weed competition as well as low input conditions such as low N with weed competition (Nissanka et al., 1996; Castleberry et al., 1984).

The determination of the physiological basis of increased productivity has been possible because open-pollinated varieties and hybrids from the 1930s are still available for examination. Stem water potential is lower in a new hybrid, Pioneer 3902 versus an older hybrid, Pride 5, and recovery following moisture stress during silking is more rapid. The canopy photosynthetic rate is higher under water stress during silking in a new hybrid versus a older hybrid. A significant finding under the same stress conditions during silking is that respiration rates in the newer hybrid are *lower* than those in an older hybrid (Nissanka et al., 1996). Phenotypic appearance also has tended toward a more upright leaf angle, which may be more efficient in light interception under high populations (Duvick, 1984, 1996b). Harvest index has increased as evidenced by reduced tassel size and branch number (Duvick, 1984, 1996b). Grain composition has been altered, unintentionally, as a result of selection. Grain protein in hybrids selected during the past 20 yr are about 1% lower in protein (10.2 vs. 9.0%, see Duvick 1992, 1996b) and are higher in starch content. Cacco et al. (1983) also have shown that S and N uptake also are more efficient in newer hybrids. The further comparison of the underlying physiological and genetic changes in modern versus older hybrids will undoubtedly set the stage for additional positive modifications at the molecular and cellular level.

The relative contribution of heterosis to the performance of hybrids from different periods from the 1940s through the 1980s has been investigated by Duvick (1984) and by Duvick and Lamkey (1996, personal communication). The heterosis of maize hybrids was uniquely positioned to take advantage of the rapid changes toward more intensive agriculture in America during the 1930s. The hybrids first described by Shull (1909) were higher yielding than the open-pollinated varieties under comparison; however, the inbred parents of these hybrids had much lower yield than their source open-pollinated varieties. Fifty years of hybrid maize breeding has increased both the yield of male and female inbred parents and the amount of heterosis observed under stress such as high population density. The relative contribution of heterosis in hybrid performance has increased over time, but the rate of increase appears to be somewhat less than the yield improvement of inbreds (Duvick, 1984; Duvick & Lamkey, 1996, personal communication). Parental inbred yields, especially female inbreds, have increased and have closely paralleled hybrid yield increases but at a lower absolute level because of inbreeding depression. The yield increases of inbred parents of maize hybrids are similar to those observed for other agronomic crops such as soybeans and small grains (Avery, 1995).

CONSEQUENCES OF ADOPTING THE TECHNOLOGY OF HETEROSIS

Hybrid maize increased productivity: the same amount of grain was produced on relatively fewer hectares. Examination of the number of hectares planted between 1927 and 1995 clearly show that fewer hectares are planted today than in decades of 1920 and 1930 (Fig. 2–3) while the total amount of harvested grain has dramatically increased. Teigen and Thomas (1995) in an extensive investigation of the impact of weather on yield from 1950 to 1994, also estimated the impact of weather and technology including genetics and farm size (see below) on grain yields of maize in Illinois and Georgia. A two-way analysis of variance was used to separate the farm size effects from year-to-year effects. The year-to-year fluctuation include both systematic (fixed) effects of technology such as genetics (e.g., hybrid adoption) fertilizer inputs (e.g., N), and management choices (e.g., increased density) and the random effects of weather, primar-

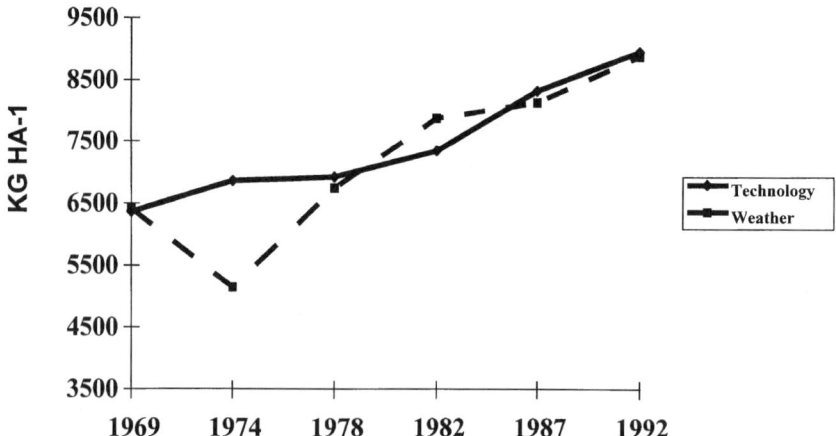

Fig. 2–4. Estimated maize yields calculated from county census data from 1969 to 1992 in Illinois counties from a two-way analysis of variance in which the simple average of yields across farm size shows the year effect expressed as a technology or weather component where yield in size s and year t is given by: $Y_{st} = \alpha_s + \mu_t + \varepsilon_{st}$, where $\Sigma s \alpha s = 0$, μ_t = weather + technology, and ε_{st} is distributed as $_{N(0, \sigma^2)}$. Source: Teigen and Thomas (1995).

ily temperature and precipitation in July and August. In general, precipitation increased yield while high temperature reduced yield. The year to year (shown as weather component in Fig. 2–4) fluctuates depending on the deviation contributed by weather, e.g., 1974 is significantly lower. The estimated year-to-year yield increases due to technology (shown as technology component in Fig. 2–4) continued to increase from 1969 to 1992 although the increase from 1974 to 1978 was small. Teigen and Thomas (1995) note that the energy crisis of 1974–1978 (less use of N fertilizers), inflation and other reasons may explain the relatively flat slope of the technology component of yield in this period.

Maize production has shifted geographically and farm size has increased in recent years. Maize and other grains have moved to areas in which climate and land quality advantages are better suited for grains (Avery, 1995). New hybrids adapted to the north central U.S. and Ontario in Canada are dramatic examples of this production shift (Fig. 2–5 and 2–6). Ontario grain maize hectares have risen dramatically, particularly since 1960, resulting in Ontario changing from a net importer to a net exporter of maize. Minnesota and South Dakota have shown increased maize hectarage and increased production from 1982 to 1992 while Virginia and West Virginia have shown declines (Fig. 2–6). Note that in each state the number of farms is either stationery, or more likely declined over the same period. The top ranking maize producing states of Iowa, Illinois, Nebraska, Indiana, and Minnesota produced two thirds of the nations maize in 1995 (USDA-NASS, 1996).

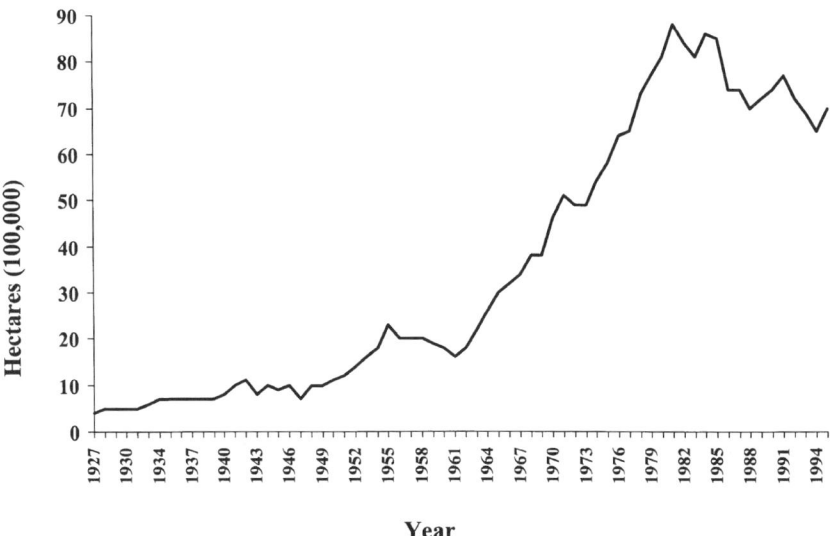

Fig. 2–5. Grain maize hectares in Ontario from 1927 to 1995. Source: Statistics Canada (1975, 1985 Ontario Corn Producers Association, 1997).

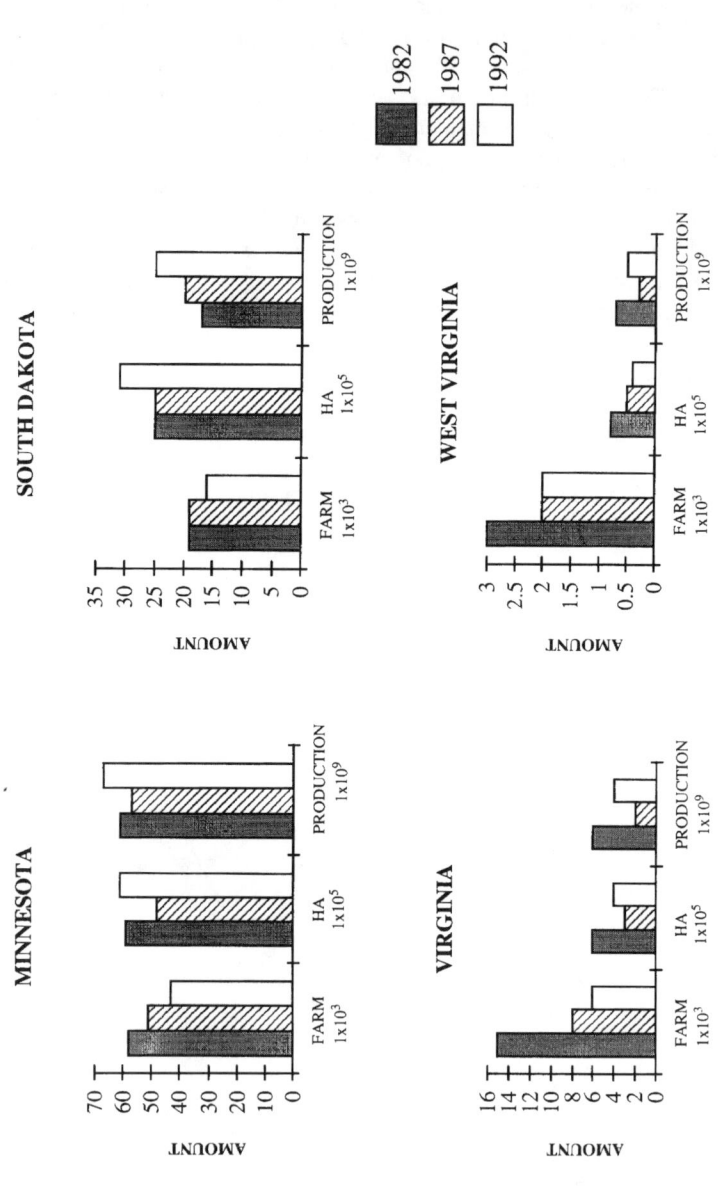

Fig. 2–6. Number of farms, grain maize hectares, and grain maize production in the states of Minnesota, South Dakota, Virginia, and West Virginia for the years of 1982, 1987, and 1992. Source: U.S. Department of Agriculture/National Agricultural Statistics Service (1996).

Table 2–2. U.S. averages of selected items for farms with sales of $100 000 or more compared with farms with sales of less than $100,000, 1992.†

Item	$100 000 or more	Less than $100 000
Land in farms (hectares)	1 542	271
Value of land and buildings (dollars)	1 059 510	212 816
Value of machinery and equipment (dollars)	150 852	27 547
Value of sales (dollars)	402 081	17 825
Grains (dollars)	125 806	19 873
Vegetables, sweet corn and melons (dollars)	332 264	10 631
Government payments (dollars)	17 171	5 080
Average net cash returns (dollars)	83 812	1 836

† Source: U.S. Department of Commerce, 1996.

EFFICIENCY: COSTS VERSUS PROFITABILITY

Another profound change applauded by some (Avery, 1995) and decried by others (Vogeler, 1981) is the growth in large farms and a reduction in the overall number of farms. The number of farms recorded in the 1995 census declined below 2 million for the first time since the Civil War (Agricultural Census, 1995). The changing pattern of farm size and number in the USA is illustrated in Fig. 2–6. Among the important issues concerning the shift to larger size farming operations are: (i) the relative costs for grain production, (ii) profitability, and (iii) the efficiency of maize production. The lowest cost farm sizes in 1967 for grain maize and vegetables was examined by Madden and Partenheimer (1972) and discussed by Vogeler (1981). The lowest cost maize grain farm in southern Iowa in 1967 was a two person farm with a three plow tractor while the lowest cost vegetable farm in the California Imperial Valley was 243 hectares with contracted services; however, Madden and Partenheimer also point out that the lowest cost farm also has the *lowest profit* with the profit increasing steadily with increasing size. Furthermore, profit continues to rise as the scale of farming operations increases while costs rise at a lower rate. Despite criticism about the higher costs of larger farms (Vogeler, 1981), profitability per farm may be the driving force behind increases in farm size (Madden & Partenheimer, 1972).

Further examination of the effect of farming operation size is summarized in Table 2–2. Farms with sales value more than $100 000 accounted for 17% of the farms and 83% of the sales. Moreover, larger scale farms accounted for 75% of grains produced and 93% of the vegetables produced in 1992. Despite the larger size, 70% of the farms with sales more than $100 000 were defined as individual or family farms (U.S. Dep. of Commerce, 1996). Thus, the trend predicted by Madden and Partenheimer (1972) has continued through the latest census data (Table 2–2). Although large farms dominate the amount of grain and vegetable production, small farms persist. Madden and Partenheimer (1972) speculate that the small farms persist because the owners sell both goods and services. Therefore, a small farm owner may work for others (services) to augment income other than that produced on the farm.

Teigen and Thomas (1995) investigated the productivity of different farm sizes for maize, soybeans [*Glycine max* (L.) Merr], and wheat (*Triticum aestivum*

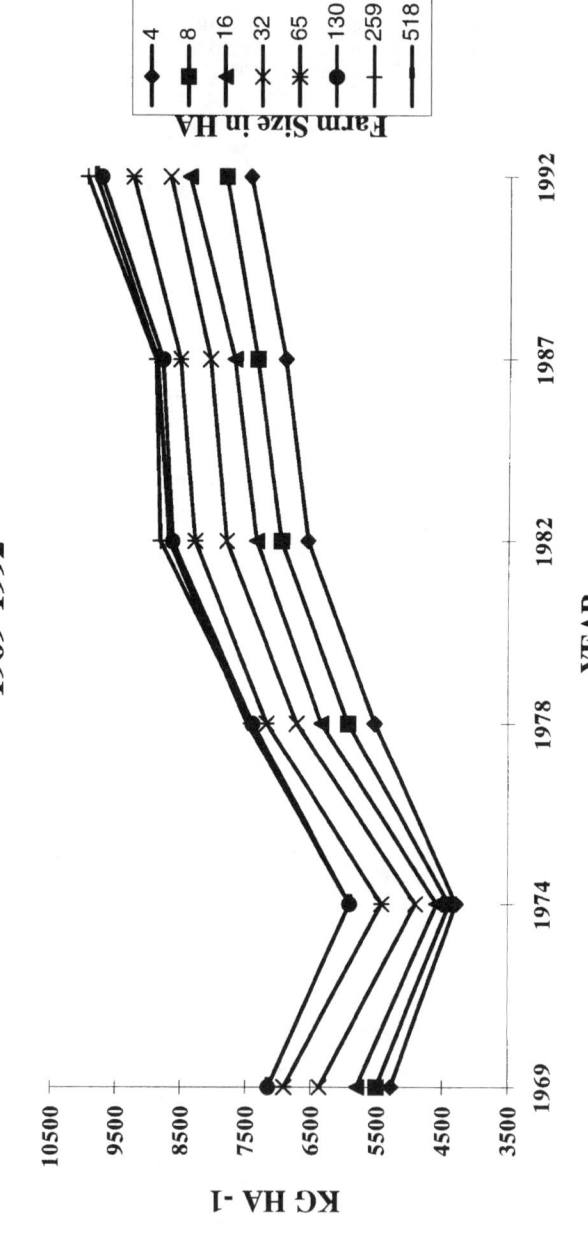

Fig. 2–7. Maize yields calculated from 1969 to 1992 in Illinois counties of farms with 4, 8, 16, 32, 65, 130, 259, and 518 hectares in Illinois from a two-way analysis of variance where yield in size s and year t is given by: $Y_{st} = \alpha_s + \mu_t + \varepsilon_{st}$, where $\Sigma S \, \alpha s = 0$, μ_t = weather + technology, and ε_{st} is distributed as $_{N(0, \sigma^2)}$. Source: Teigen and Thomas (1995).

Table 2–3. Fresh market tomato area (ha) planted, harvested yield in thousands cwt, price/cwt, and value in years 1992, 1993, 1994, and 1995.†

Year	Planted	Harvested	Ton	Price
	ha		million	
1992	56 571	53 382	1.77	$35.80
1993	56 571	53 382	1.61	$35.80
1994	55 191	53 670	1.66	$27.50
1995	55 001	53 305	1.49	$26.00

† Source: USDA–National Agricultural Statistics Service, 1996.

L.) using the county census data in Illinois (see above) and other states with the data for maize in Illinois shown in Fig. 2–7. Examination of data on a state basis eliminates bias because of regional differences in productivity. Larger farms are clearly more productive than smaller farms, with the gap between small and large farms widening as yield has increased between 1969 and 1992. Approximately 2 T ha^{-1} separate the highest and lowest yielding farms in Illinois. A similar response also was observed in Georgia (data not shown); however, the highest maize yields in Illinois in 1992 were in the next to largest size class rather than the largest size. In summary, larger farms offer higher profits per farm, produce an increasing proportion of grain maize, and vegetables, and in general produce higher yields than smaller farms.

While maize grain is regarded as a quantitative trait, long shelf life in tomatoes is an economic yield factor conditioned by both quantitative and single genes present in hybrids (heterozygous, see below for explanation). Elsberry (1995) demonstrated that extended shelf-life tomatoes could be grown and marketed, depending on price, to increase income to the grower–packer in the Homestead, FL, area because of time flexibility for harvest. Since this trait might be considered as a gene for yield, the eventual impact in terms of hectares and locations are unclear. Recent statistics from the U.S. agricultural census (USDA-NASS, 1996) suggest that hectares grown and yields have been stable for fresh market tomatoes (Table 2–3). Price has clearly declined, albeit the underlying reasons are somewhat unclear. Although adoption of longer shelf life hybrids might be one explanation (thus giving a greater available supply), changes in import rules also have a strong impact on supply and price.

HYBRIDS: HETEROSIS PLUS

Heterosis for grain yield has been very well documented (see other contributions from this series), but also represents only one component of performance that a specific combination of parents contribute. Another example of hybrids is fresh market tomatoes. Tomato shelf life illustrates the importance of specific parental contributions to an important economic trait. Fresh market tomatoes generally have a shelf life of about 5 to 7 d from the time the fruit is at the red-ripe stage. Commercially, fruit is picked at the mature green stage, stored, and then treated with ethylene to allow slow ripening during transportation to market destinations distant from the major growing areas. Fruit of plants carrying two recessive mutations *rin* and *nor*, do not complete normal ripening, but remain in an intermediate state for a period of months. The *rin* and *nor* mutants at 70 d after pollination have the same firmness as normal Rutgers fruit at 43 d (Mitcham et al., 1991). The use of either *rin* and *nor* as a male parent results in hybrids in which shelf life is increased by 14 to 28 d. Tomatoes can be picked at the red ripe stage, stored by packers and then directly transported to markets in contrast to the traditional approach of picking at the mature green stage followed by storage and treatment with ethylene. The shelf life trait can be viewed as a yield enhancer since it allows producers to spread the time period in which tomatoes can be harvested from a given date of planting.

PRIVATE AND PUBLIC RESEARCH FUNDING

Hybrid maize serves as a unique example in which the technology of heterosis for grain yield effectively delivered a superior product into the hands of farmers. Hybrids offered a ready means in which the farmers who used the hybrids became more profitable and the companies selling hybrids also were more profitable (Duvick, 1996a). The American hybrid seed maize industry thrived because of these incentives for development of private sector research, seed production and seed distribution. The contribution of both public and private research have been well documented (Duvick, 1996a), and suggest a continuing need for research that contributes to products that reach the marketplace. Customer demand and the desire to obtain commercially successful maize hybrids determine research direction and the continual evolution of new technology, especially in private companies. Further, the scale of research effort depends not only on the current market share of a specific company, but also on the growth potential afforded by customer acceptance of new products derived from an investment in research (Duvick, 1996a).

REFERENCES

Avery, D.T. 1995. Seeds of success. P. 237–253. *In* Saving the planet with pesticides and plastic: The environmental triumph of high-yield farming. Hudson Inst., Indianapolis, IN.
Ball, A.G., and E.O. Heady. 1972. Trends in farm and enterprise size and scale. p. 40–58. *In* A.G. Ball and E.O. Heady (ed.) Size, structure and future of farms. Iowa State Univ. Press, Ames.
Cacco, G., M. Saccomani, and G. Ferrari. 1983. Changes in the uptake and assimilation efficiency for sulfate and nitrate in maize hybrids selected during the period 1930 through 1975. Physiol. Plant 58:171–174.
Castleberry, R.M., C.W. Crum, and C.F. Krull. 1984. Genetic yield improvement of U.S. maize cultivars under varying fertility and climatic environments. Crop Sci. 24:33–36.
Cavalieri, A.J., and O.S. Smith. 1984. Grain filling and field drying of a set of maize hybrids released from 1930 to 1982. Crop Sci. 25:856–860.
Duvick, D.N. 1984. Genetic contributions to yield gains of U.S. hybrid maize, 1930–1980. p. 15–47. *In* D.E. Weibel et al. (ed.) Genetic contributions to yield gains of five major crop plants. CSSA Spec. Publ. 7. CSSA, Madison, WI.
Duvick, D.N. 1992. Genetic contributions to advances in yield of USA maize. Maydica 37:69–79.
Duvick, D.N. 1996a. The United States. p. 193–212. *In* M. Morris (ed.) Maize seed industries in developing countries: Technical, economic, and policy issues. CIMMYT, El Batan, Mexico.

Duvick, D.N. 1996b. What is yield? p. 332–335. *In* G.O. Edmeades (ed.) Developing drought and low N-tolerant maize. CIMMYT, El Batan, Mexico.

Elsberry, E, 1995. Income comparison between regular and super life tomatoes grown and harvested as mature greens. Pioneer Vegetable Genetics Marketing Brochure, Pioneer Hi-Bred International, Des Moines, IA.

Madden, J.P., and E.J. Partenheimer. 1972. Evidence of economies and diseconomies of farm size. p. 91–107. *In* A.G. Ball and E.O. Heady (ed.) Size, structure and future of farms. Iowa State Univ. Press, Ames.

Mitcham, E.J., K.C. Gross, and T.J. Ng. 1991. Ripening and cell wall synthesis in normal and mutant tomato fruit. Phytochemistry 30(6):1777–1780.

Nissanka, S.P., M.A. Dixon, and M. Tollenaar. 1996. Canopy gas exchange response to moisture stress in old and new maize hybrid. Crop Sci. 37:172–181.

National Corn Growers Association. 1995. The Worlds of Corn 1996. Natl. Corn Growers Assoc., St. Louis, MO.

Shull, G.H. 1909. A pure-line method in corn breeding. Am. Breeders Assoc. 5:51–59.

Statistics Canada. 1975. Handbook of agricultural statistics. Part 1. Field crops, 1921–1974. Canadian Gov. Publ., Ottawa, ON.

Statistics Canada. 1985. Handbook of field crop area, yield, production, and average farm price and value, 1953–1983. Nov 1985. Canadian Gov. Publ., Ottawa, Canada.

Ontario Corn Producers Association. 1997. Ontario Corn Producers Association facts and figures. Guelph, ON.

Teigen, L.D., and M. Thomas, Jr. 1995. Weather and yield, 1950–94: Relationships, distributions, and data. Commercial Agric. Div., Economic Res. Serv., USDA Staff Pap. 9527. U.S. Gov. Print. Office, Washington, DC.

Tollenaar, M. 1992. Is low plant density a stress in maize? Maydica 37:305–311.

Tollenaar, M., A. Aguilera, and S.P. Nissanka. 1997. Grain yield is reduced more by weed interference in an old than in a New Maize Hybrid. Agron. J. 89:239–246.

Tollenaar, M., L.M. Dwyer, and D.W. Stewart. 1992. Ear and kernel formation in maize hybrids representing three decades of grain yield improvement in Ontario. Crop Sci. 32:432–438.

Vogeler, I. 1981. The myth of the family farm: Agribusiness dominance of U.S. agriculture. Westview Press, Boulder, CO.

U.S. Department of Agriculture/National Agriculture Statistics Service. 1996. Statistical highlights of U.S. agriculture, 1995/96. USDA-NASS, Washington DC.

U.S. Department of Agriculture. 1969. USDA Statistics Bull. 233. U.S. Gov. Print. Office, Washington, DC.

U.S. Department of Commerce. 1996. Large farms are thriving in the U.S. USDC Bureau of the Census Agricultural Brief AB/96-1.

3 Overview of Heterosis and Heterotic Groups in Agronomic Crops

Albrecht E. Melchinger and Ramesh K. Gumber

*Institute of Plant Breeding, Seed Science
 and Population Genetics
University of Hohenheim
Stuttgart, Germany*

ABSTRACT

Heterotic groups and patterns are of fundamental importance in hybrid breeding. We start with definitions of these terms. Theoretical and experimental arguments are given demonstrating the superiority of inter-group compared with intra-group crosses under two aspects: (i) a higher mean heterosis and hybrid performance and (ii) a reduced specific combining ability (SCA) variance and lower ratio of SCA to general combining ability (GCA) variance, which implies that identification of superior hybrids can be based mainly on testing for GCA. We review the degree of heterosis, history and current status of hybrid breeding, and development of heterotic groups in five major crops with different pollination systems: allogamous—maize (*Zea mays* L.) and rye (*Secale cereale* L.); partially allogamous—faba bean (*Vicia faba* L.) and oilseed rape (*Brassica napus* L.); autogamous—rice (*Oryza sativa* L.). Fundamental principles and systematic approaches for identification of heterotic groups and patterns and enlarging their genetic base are suggested with special consideration of the use of molecular markers for grouping of germplasm. Adapted populations, isolated either by time and/or space are most suitable candidates for promising heterotic patterns. The potential of heterotic groups in clone and population breeding also is discussed.

Schnell (1982) pointed out that heterosis is a major yield factor in all breeding categories except line breeding; however, only in hybrid and clone breeding it is possible to have maximum exploitation of heterosis. High heterosis can be expected from a hybrid if the source populations have (i) a high frequency of partially or completely dominant genes, and/or (ii) maximum difference in gene frequencies for overdominant loci. Consequently, for optimum exploitation of heterosis in hybrid breeding, the seed and pollen parents should be derived from genetically unrelated germplasm pools, commonly referred to as heterotic groups.

Copyright © 1998 Crop Science Society of America, 677 S. Segoe Rd., Madison, WI 53711, USA.
Concepts and Breeding of Heterosis in Crop Plants. CSSA Special Publication no. 25.

Table 3–1. Estimates of midparent heterosis (in percentage of mid-parent performance) for grain yield in different crops.

Crop/mating system	Heterosis			Reference
	Mean	Minimum	Maximum	
		%		
Allogamous				
Maize—U.S.	121	92	240	Dudley et al., 1991
—Europe	129	112	143	Melchinger et al., 1986
Rye	178	86	301	Geiger & Schnell, 1975
	207	117	329	Geiger & Wahle, 1978
Partially allogamous				
Faba bean	45	22	69	Kittlitz, 1986
	74	55	95	Link et al., 1996
Oilseed rape—spring	30	20	50	Grant & Beversdorf, 1985
—winter	50	20	80	Lefort-Buson & Dattee, 1982
Autogamous				
Rice	36	3	106	Saghai Maroof et al., 1997
	55	31	73	Virmani et al., 1982
Wheat	9	−14	106	Martin et al., 1995

In our terminology, a heterotic group denotes a group of related or unrelated genotypes from the same or different populations, which display similar combining ability and heterotic response when crossed with genotypes from other genetically distinct germplasm groups. By comparison, the term heterotic pattern used herein refers to a specific pair of two heterotic groups, which express high heterosis and consequently high hybrid performance in their cross.

Heterotic groups have a strong impact in crop improvement because they predetermine to a large extent the type of germplasm used in a hybrid breeding program over a long period of time. The question of suitable heterotic groups and patterns is becoming relevant in many crops, where hybrid breeding has or will become feasible with the development of new genetically-engineered systems of pollination control (Williams, 1995).

The objectives of this review are to describe (i) the importance of heterotic groups in hybrid breeding, (ii) degree of heterosis, (iii) history and current status of hybrid breeding, and (iv) development of heterotic groups in five selected crops. We also suggest fundamental principles and systematic approaches that might be used for the identification of heterotic groups and patterns as well as the enlargement of their genetic base.

REVIEW OF HETEROSIS IN DIFFERENT CROPS

Throughout this chapter, the term heterosis refers to the deviation of the F_1 hybrid from the mean ($\bar{P}$) of two homozygous parent lines. Analogously, the term heterotic deviation denotes the difference of the population cross from the means of two parental populations. Table 3–1 summarizes estimates of heterosis for grain or seed yield, expressed as percentage of $\bar{P}$, for selected crops from various studies in the literature. In crosses between elite lines of maize (*Zea mays* L.; Dudley et al., 1991; Melchinger et al., 1986), mean heterosis for grain yield was of similar magnitude (~125%) in germplasm from the U.S. Corn Belt and Europe. In a study with rye (*Secale cereale* L.; Geiger & Schnell, 1975; Geiger & Wahle, 1978) using first-cycle inbred lines derived from open-pollinated varieties, mean heterosis was extremely large (~200%) due to the low yield potential of the parents. Among partially allogamous crops, mean heterosis for yield was similar in faba bean (50–60%; Kittlitz, 1986; Link et al., 1996) and oilseed rape (30–50%; Grant & Beversdorf, 1985). Despite its autogamous nature, a substantial mean heterosis for yield (35–55%) was reported in rice (Saghai Maroof et al., 1997; Virmani et al., 1982). However, heterosis for seed yield was only 9% in wheat (*Tritcum aestivum* L.; Martin et al., 1995). These studies clearly demonstrate that heterosis for grain or seed yield is usually highest for allogamous crops and considerably lower for partially allogamous and autogamous species.

EXPERIMENTAL ARGUMENTS SUPPORTING THE CONCEPT OF HETEROTIC PATTERNS

Means of Intra-Group vs. Inter-Group Hybrids

The superiority of inter-group over intra-group crosses in terms of mean performance and heterosis or heterotic deviation for grain yield and other heterotic traits has been well documented in the literature. In maize, with genetically balanced sets of crosses, inter-group hybrids outyielded the respective intra-group hybrids by 21% in Reid Yellow Dent (RYD) × Lancaster Sure Crop (LSC) crosses (Dudley et al., 1991) and by 16% in Flint × Dent crosses (Dhillon et al., 1993) (Fig. 3–1). In both studies, the percentage of increase in heterosis for yield of inter-group over intra-group crosses was about twice as large as for hybrid yield itself. In rice, inter-group hybrids between *O. indica* and *O. japonica* derived lines displayed substantially greater heterosis and F_1 yield than intra-group crosses (Xiao et al., 1996). Compared with intra-group crosses, the parental genetic distance of inter-group crosses, calculated from RFLP or RAPD data, was only moderately increased in maize, but substantially higher in rice (Fig. 3–1).

In Fig. 3–2, we compared the performance of inter-population and intra-population crosses, the latter corresponds to the per se performance of the populations. In two studies with maize germplasm from the U.S. Corn Belt, inter-population crosses outyielded intra-population crosses by 16 to 24% with an average positive heterotic deviation of about 20% (Kauffmann et al., 1982; Mungoma & Pollak, 1988). In rye, crosses among populations from different germplasm pools such as Petkus and Carsten had on average 11% higher yield than intra-pool population crosses (Hepting, 1978). In spring oilseed rape, two inter-population crosses between open-pollinated cultivars of Canadian and European origin outyielded their parent populations by 50%, the average heterotic deviation (ΔH) across all inter-population crosses amounted to 16% (Grant & Beversdorf, 1985).

These results indicate that heterosis or heterotic deviation increases with an increase in genetic distance between the parents; however, Moll et al. (1965)

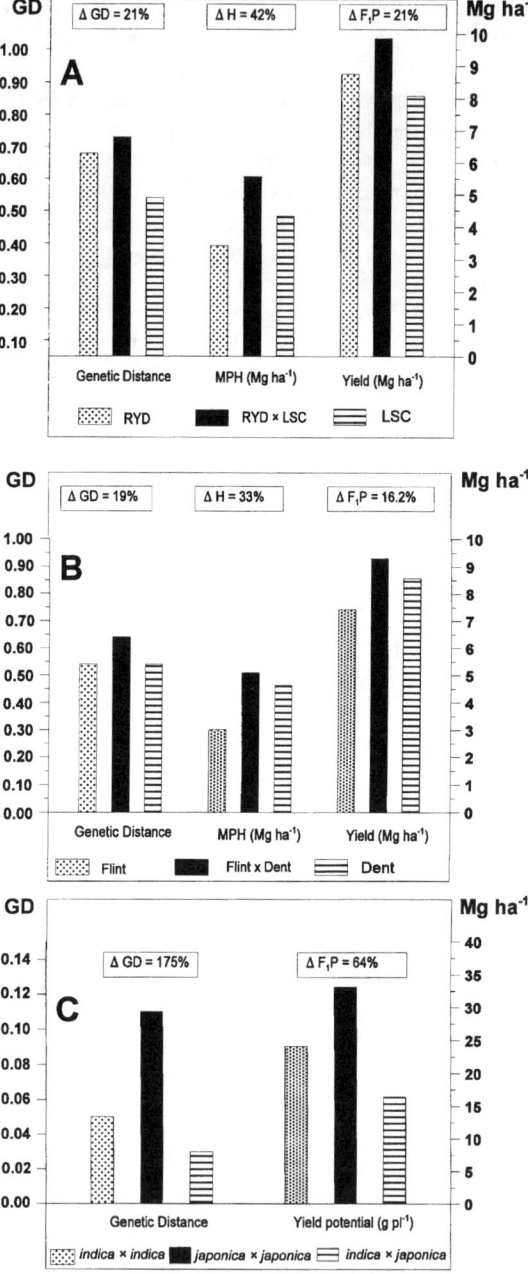

Fig. 3–1. Comparison of genetic distance, midparent heterosis, and yield of intra- and inter-group hybrids in (A) U.S. maize, (B) European maize, and (C) rice. Data taken from Dudley et al. (1991), Dhillon et al. (1993), and Xiao et al. (1996), respectively. ΔH, ΔGD, and $\Delta F_1 P$ denote the deviations for heterosis, genetic distance, and F_1 performance, respectively, in cross $\pi_1 \times \pi_2$ calculated as: $\Delta = [\ 100 * \pi_1 \times \pi_2 / (\ \pi_1 \times \pi_1 + \pi_2 \times \pi_2) / 2\ ] - 100$, where π_1 and π_2 refers to the two populations investigated.

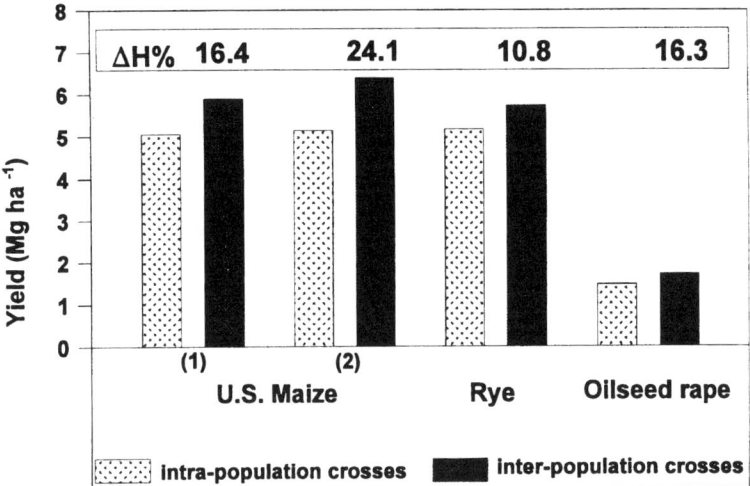

Fig. 3–2. Comparison of intra- and inter-population crosses for yield in U.S. maize, rye, and oilseed rape. Data taken from (1) Kauffmann et al. (1982), (2) Mungoma and Pollak (1988) in U.S. maize, Hepting (1978) in rye, and Grant and Beversdorf (1985) in oilseed rape.

demonstrated that this relationship holds true only up to a certain optimum level of genetic distance and thereafter, one finds a decrease in hybrid performance and heterosis as depicted schematically in Fig. 3–3. Link et al. (1996) stressed in this context that hybrid performance and heterosis are not merely a function of genetic distance but also depend on the adaptation of the parents.

Variances Among Intra-Group vs. Inter-Group Hybrids

The variances due to general (σ^2_{GCA}) and specific combining ability (σ^2_{SCA}) and their ratio are very important for predicting hybrid performance (Melchinger et al., 1987). With predominance of σ^2_{GCA} over σ^2_{SCA}, early testing becomes more effective and superior hybrids can be identified and selected mainly based on their prediction from GCA effects. We reanalyzed the data of a diallel study with

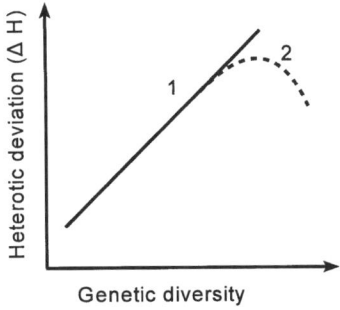

Fig. 3–3. Schematic relationship between heterosis in varietal crosses and genetic diversity of their parents based on data from (1) Moll et al. (1962) and (2) Moll et al. (1965) in maize.

Table 3–1. Estimates of variance components due to general combining ability (GCA) and specific combining ability (SCA) in inter-group crosses (IGC) and in diallel crosses without considering heterotic groups (WCG) for different forage traits in maize (Melchinger, unpublished results, original data from Dhillon et al., 1990).

Trait	IGC			WGC		
	σ^2_{GCA}	σ^2_{SCA}	$\sigma^2_{SCA}:\sigma^2_{GCA}$	σ^2_{GCA}	σ^2_{SCA}	$\sigma^2_{SCA}:\sigma^2_{GCA}$
			kg ha^{-1}			
Total dry matter yield	77.68	38.94	0.50	49.93	194.03	3.89
Ear dry matter yield	17.51	5.34	0.30	5.24	67.89	12.94
Stover dry matter yield	33.95	19.43	0.57	31.50	36.57	1.16
Mean ratio			0.49			3.45

six Flint and six Dent inbreds published by Dhillon et al. (1990). Estimates of σ^2_{SCA} and ratio $\sigma^2_{SCA} : \sigma^2_{GCA}$ were much lower for grain and forage yield in the factorial of inter-group crosses (Flint × Dent) than in the complete diallel, treating the Flint and Dent inbreds as a sample from one common parent population (Table 3–2). These findings are in agreement with theoretical results indicating that inter-group crosses have smaller σ^2_{SCA} and $\sigma^2_{SCA} : \sigma^2_{GCA}$ ratios than intra-group crosses (Melchinger, 1996, unpublished data).

STATUS OF HYBRID BREEDING AND HETEROTIC GROUPS IN DIFFERENT CROPS

The experimental results clearly demonstrated the superiority of inter-group over intra-group hybrids and the necessity of heterotic groups in hybrid breeding. Despite their great importance, most of the primary heterotic groups were not developed in a systematic manner. They were determined empirically by relating the observed heterosis and hybrid performance with the origin of parents included in the crosses. In this section, we review the development of primary heterotic groups and patterns in five selected crops.

Maize

U.S. Corn Belt and Canada

Richey (1922) provided the first important information that eventually led to the development of heterotic patterns in maize. He compared heterosis for yield in a large number of inter-varietal crosses and found higher performance for hybrids between varieties of different endosperm type than those among varieties possessing the same endosperm type. His investigation strongly suggested that crosses between geographically or genetically distant parents express higher performance and consequently increased heterosis.

Based on this principle, the most widely used heterotic pattern in the U.S. Corn Belt, RYD × LSC, was evolved. The two parent populations of this heterotic pattern were adapted to different geographic regions and also had different genetic backgrounds (Hallauer et al., 1988). The variety RYD was originated in Western Illinois by a chance hybridization between strains 'Gordon Hopkins' and

'Little Yellow', while the open-pollinated variety LSC was developed from the cross of an early Flint population with a late local variety in Lancaster County, Pennsylvania.

A first systematic attempt to identify alternative heterotic patterns to RYD × LSC was reported by Kauffmann et al. (1982). They evaluated diallel crosses among 10 U.S. Corn Belt open-pollinated varieties and identified two promising alternative heterotic groups: Leaming and Midland. These groups were originated in different regions of the U.S. Corn Belt and have distinct morphological characteristics: Leaming developed in southern Ohio has a slender ear, while Midland from southern Kansas has a wider leaf plant type, short ear, and more kernel rows. They combine well with each other and with LSC (Hallauer et al., 1988). These two new heterotic groups were identified in an advanced breeding stage and thus they were not competitive with germplasm derived from the well-established RYD and LSC heterotic groups, which had undergone numerous selection cycles for agronomic traits. More recently, Mungoma and Pollak (1988) evaluated diallel crosses among seven yellow and three white endosperm type populations to identify new heterotic patterns. They suggested that a Mexican Dent population can be used as an alternative to LSC in the U.S. Corn Belt heterotic pattern.

Central and Western Europe

When hybrid breeding was started in Europe after World War II, attempts were made to introduce early-maturing Dent hybrids from Canada and the U.S. Corn Belt; however, these were not sufficiently adapted to the cooler climatic conditions of Central Europe (Schnell, 1992). In the 1950s, indigenous Flint populations such as Lizagaraute, Lacaune (both from France) as well as Gelber Badischer Landmais and Rheintaler (from Germany) became the ancestors of the current Flint germplasm. The productivity of Flint lines like F2, F7, EP1, and DK 105 derived from these ancestral populations was generally low. Therefore, the Flint lines were crossed to early-maturing highly productive Dent lines of North American origin to develop high yielding adapted hybrids. This established the complementary Flint × Dent heterotic pattern.

The source populations of the Dent lines were derived from the cross of 'early slender-stalked' Flint corn of the north-eastern USA with the 'late heavy stalked gourd seed' of the south-central USA. The Flint populations can be traced back to tropical Flints originating from the West Indies and the Caribbean Islands (Wallace & Brown, 1956) and had been isolated from Dent populations for at least 500 yr.

Tropics and Subtropics

Promising heterotic patterns used in the tropics are ETO × Tuxpeno, Suwan I × Tuxpeno, Cuban Flints × Tuxpeno, and Caribbean Flints × Tuxpeno (M. Bjarnason, 1996, personal communication). In addition, Caribbean Flints × Tuxpeno in Brazil, Mexico, and Colombia, and ETO × Caribbean Flints in Venezuela, and American early Dent × Tuxpeno in Egypt also are used in hybrid breeding (Eberhart et al., 1995; Pandey & Gardner, 1992).

A systematic search by Goodman (1985) resulted in the identification of 10 heterotic patterns based on testcross evaluation of tropical races. He suggested the use of only those promising heterotic patterns that involved populations of Tuxpeno, Tuson, Cuban Flint, Coastal tropical Flint, and Chandelle. To identify heterotic patterns among CIMMYT quality protein maize (QPM) germplasm, Vasal et al. (1993) evaluated diallel crosses among 10 parents, six QPM gene pools and four QPM populations. They suggested Pool 32 QPM, Population 68 QPM, and Population 69 QPM as ideal candidates for promising heterotic patterns due to their high grain yield and good combining ability.

Rye

Hybrid breeding in rye was started with the detection of cytoplasmic-genic male sterility (Pampa CMS) from the Argentinean population 'Pampa rye' by Geiger and Schnell (1970) more than two decades ago. First commercial hybrids were released in 1985. They had 15 to 20% higher grain yield but slightly lower plant height than open-pollinated population varieties. Presently >50% of the rye area in Germany is planted with hybrids.

The systematic search for suitable heterotic patterns by Hepting (1978) laid the foundation of hybrid breeding in rye. Based on the hybrid performance and heterotic deviation in a complete diallel among seven open-pollinated populations, he found that cross combinations involving populations from the two most widely used germplasm groups, Petkus and Carsten, were most promising for grain yield. In fact, all rye hybrids released in Germany since 1985 are based on the Pampa CMS system and are of the Petkus × Carsten type (H.H. Geiger, 1996, personal communication).

The Petkus and Carsten germplasm groups originated from two open-pollinated populations and were developed from different landraces. Both populations were well adapted to temperate continental climate but different soil conditions. The use of different source materials and independent selection for different plant ideotypes resulted in genetic differences for various traits. Besides their different history of evolution, they were separated from each other for at least 100 yr. Allozyme data confirmed a greater genetic distance between the Petkus and Carsten germplasm groups than within Petkus and Carsten (Adam et al., 1987). Due to its wide adaptation and broad genetic base, the Petkus material has been used either directly or as an important germplasm source in most rye breeding programs in the world. Carsten germplasm has high lodging resistance and greater number of kernels per ear but a smaller 100-kernel weight than Petkus (H.H. Geiger, 1996, personal communication).

Oilseed Rape

The presence of substantial heterosis for seed yield (up to 72%) and effective pollination control systems (Ogura, 1968; Williams, 1995) stimulated the research on hybrid development in oilseed rape. Recently, commercial hybrids have been introduced in Canada, USA, France, and Germany. Hybrids have 15 to 22% higher seed yield and improved yield stability than the best line varieties.

No distinct heterotic groups comparable to maize have been established so far in winter and spring oilseed rape. Winter and spring cultivars have been cultivated in different geographical regions and represent distinct germplasm groups with little inter-crossing in the past (Diers & Osborn, 1994). Spring cultivars are grown extensively in Canada, Australia, and Scandinavia, while winter cultivars are grown in the rest of Europe and Japan.

In a search for suitable heterotic patterns, Grant and Beversdorf (1985) evaluated diallel crosses among six open-pollinated cultivars of European and Canadian spring rapeseed. Based on hybrid performance and heterotic deviation, they suggested European spring × Canadian spring rapeseed as a promising heterotic pattern. Similarly, Lefort-Buson et al. (1987) found crosses between winter rapeseed of European × Asian origin to be most productive. The results of diallel crosses among seven spring rapeseed cultivars of European, Canadian, and Asian origin further confirmed that crosses between European × Asian and Canadian × Asian spring oilseed rape exhibit higher heterosis than crosses between parents originating from the same region (Brandle & McVetty, 1990). More recently, Diers and Osborn (1994) and Becker et al. (1995) classified winter, spring, and Asian (both spring and winter) oilseed rape into three different groups using RFLP assays.

The potential of heterosis between winter and spring oilseed rape crosses has not been investigated so far. A recent proposal (T. Osborn, 1993, personal communication) has suggested the development of separate complementary heterotic groups for winter and spring oilseed rape by crossing genetically diverse lines of winter and spring rapeseed. Selection should be performed in subsequent backcross generations for plants having most of their genome from winter rapeseed but with spring growth habit. A reciprocal procedure could be applied to select for the genetic background of spring oilseed rape but with vernalization requirements and frost resistance; however, the latter may be more difficult because frost resistance and winter hardiness are quantitative traits.

Krahling (1987) suggested the use of resynthesized lines to develop new genetically divergent groups and to widen the genetic base of spring and winter oilseed rape. Engqvist and Becker (1994) recommended first improving the resynthesized lines by recurrent selection before using them in hybrid breeding with winter and spring oilseed rape. Moreover, the genetic distance of resynthesized lines to current rapeseed germplasm needs to be checked as many of the resynthesized lines showed a rather small genetic distance to winter and spring rapeseed lines measured by RFLPs (Becker et al., 1995).

Faba Bean

Populations or open-pollinated varieties are presently under cultivation in faba bean. Due to the partially allogamous propagation of the crop, these varieties use only about half of the available heterosis. Hybrid breeding in faba bean would be very promising because it allows maximum exploitation of heterosis and combines high yield potential with improved yield stability, a major problem in this crop (Stelling et al., 1994); however, no hybrid has been released on a commercial scale due to complications associated with the existing hybridization systems. The discovery of two new CMS systems and effective restorers have opened new avenues for hybrid breeding (Link et al., 1997).

If hybrid breeding becomes feasible in faba bean, the following three well-established germplasm pools are available: Minor, Major, and Mediterranean. Minor germplasm are tall (up to 180 cm), small-seeded spring beans, mainly used for animal feed. Major germplasm are generally shorter (up to 120 cm), large-seeded spring beans, cultivated primarily for direct human consumption. Mediterranean germplasm has a short stature (up to 110 cm) and comprises small, medium-, and large-seeded winter beans. The Minor and Major groups are grown in Central and North-western Europe, while the Mediterranean group is cultivated in southern Europe and northern Africa.

Link et al. (1996) conducted a systematic search for the most promising heterotic pattern among these three heterotic groups for two mega-environments: Central Europe and the Mediterranean region. Based on the F_1 performance and mid-parent heterosis between balanced sets of crosses, the authors suggested the Minor × Major heterotic pattern for Central European environments. The Minor × Mediterranean heterotic pattern also has great promise and could be an alternative, if lodging is prevented (Link et al., 1996). Based on RAPD assays (Link et al., 1995), the large genetic distance between the Minor and Mediterranean groups is responsible for the expression of high heterosis and hybrid performance. For the Mediterranean environments, intra-group crosses of Mediterranean germplasm was recommended as a short-term solution for hybrid breeding. The Minor × Mediterranean heterotic pattern seems to be a suitable alternative but the extremely poor adaptation of Minor germplasm to Mediterranean environments may hamper their direct use (Kittlitz et al., 1993). Because the genetic distance, measured with RAPD markers, within the Mediterranean group is comparatively high (Link et al., 1995), it might be possible to identify sufficiently divergent heterotic groups within the Mediterranean germplasm.

Rice

Substantial heterosis for grain yield, first reported by Jones (1926), and an effective cytoplasmic male sterility system ('WA' CMS; Lin & Yuan, 1980) promoted hybrid breeding in rice. Hybrid rice has contributed significantly to the dramatic increase of rice production in the world. In China alone, hybrid rice is grown annually on about 17 million ha, which corresponds to 55% of the total area under rice cultivation (Xiao et al., 1995). New hybrid rice varieties have a 30 to 45% yield advantage over conventional line cultivars (Yuan, 1992). Encouraged by the Chinese success in hybrid rice, several other countries in Asia as well as Brazil, Mexico, and the USA have initiated hybrid rice breeding programs. High yielding hybrids have been produced from crosses between *O. indica* and *O. japonica* (Xiao et al., 1995); however, hybrids between *indica* and *japonica* rices show a variable degree of sterility. Moreover, fertility restorer frequency is lower in *japonica* rice than *indica* rice. Presently, hybrids involving *indica* × *indica* crosses are under cultivation and express as much as 70% heterosis.

The three subspecies of *O. sativa* (i.e., *indica*, *japonica*, and *javanica*) have different morphological and physiological characteristics and ecogeographical distribution and, therefore, may serve as distinct heterotic groups. Genotypes belonging to *indica* and *japonica* are adapted to tropical and temperate climates, respectively, whereas *javanica* rice has intermediate characteristics. The average genetic distance measured by RFLPs between *indica* lines was three to four times higher than that between *japonica* lines (Zhang et al., 1992). Yuan (1992) pointed out that heterosis for grain yield in crosses among the three subspecies has the following trend: *indica* × *japonica* > *indica* × *javanica* > *javanica* × *japonica* > *indica* × *indica* > *japonica* × *japonica*. This is in harmony with the increased genetic distance between parents of *indica* × *japonica* crosses compared with *indica* × *indica* and *japonica* × *japonica* crosses found in a recent study with RAPD markers (Xiao et al., 1996).

Zhang et al. (1995) studied molecular divergence and hybrid performance in diallel crosses among eight *O. indica* lines representing the parents of the best performing commercial rice hybrids grown in China. Their results suggest the existence of two heterotic groups within *O. indica*, one comprising rice strains from southeast China and the other comprising strains from Southeast Asia.

CRITERIA FOR THE IDENTIFICATION OF NEW HETEROTIC GROUPS AND PATTERNS

The above review indicates that heterotic groups and patterns in many crops were identified by relating the heterosis observed in crosses with the origin of the parents included in the crosses. We recommend the following criteria for the identification of heterotic groups and patterns in descending order of importance:

1. high mean performance and large genetic variance in the hybrid population to ascertain future selection response by adopting the usefulness concept of Schnell (1978);
2. high per se performance and good adaptation of both or at least one of the parental heterotic groups;
3. low inbreeding depression in the source materials for the development of inbreds; and
4. a stable CMS system without deleterious side effects, as well as effective restorers and maintainers, if hybrid breeding is based on cytoplasmic male sterility.

While the mean hybrid performance of a large number of crosses can be determined rather accurately by replicated field evaluation, estimation of the genetic variation within parent or hybrid populations by field evaluation of a large number of S_1, half-sib, or full-sib progenies per population is very laborious and resource demanding. Alternatively, molecular markers such as RFLPs, AFLPs, or RAPDs could be employed, which have proved very effective in measuring the genetic diversity at the DNA level (Melchinger, 1993).

APPROACHES FOR THE IDENTIFICATION OF HETEROTIC GROUPS AND PATTERNS

The fundamental principle underlying the approaches for the identification of heterotic patterns is to select, from a large number of germplasm sources, parent populations of crosses with highest hybrid performance. Diallel and testcrosses are generally used for this purpose. The various steps involved in the identification of heterotic groups and patterns depend on the source materials and the availability of established heterotic patterns. In principle, these approaches could also be used for enlarging the genetic base of existing heterotic groups.

Small Number of Populations

We suggest producing a complete diallel with a small number of populations. The diallel crosses together with their parent populations are evaluated in replicated field trials for hybrid performance and heterotic deviation. Using the identification criteria given above, parent populations of cross combinations with high performance are selected as potential heterotic groups and patterns. If well-established heterotic patterns are available, their performance is compared with the new heterotic patterns.

New heterotic patterns in the U.S. Corn Belt such as Midland × Leaming and Midland × LSC (Kauffmann et al., 1982) or BSSS(R)C10 × Mexican Dent (Mungoma & Pollak, 1988) were identified by this approach. Similarly, Hepting (1978) in rye and Grant and Beversdorf (1985), Lefort-Buson et al. (1987) and Brandle and McVetty (1990) in oilseed rape identified promising heterotic patterns using diallel crosses of open-pollinated populations.

Large Number of Germplasm Accessions

Case 1: Heterotic Patterns Are Not Available

With large number of inbred lines or populations available, it is not feasible in most crops to make diallel crosses and produce sufficient F_1 seed for multi-environment field testing. Therefore, we suggest a multi-stage procedure to identify heterotic groups, which consists of the following steps:

1. grouping the germplasm based on genetic similarity;
2. selection of representatives genotypes (e.g., two to four lines or one population) from each subgroup for producing diallel crosses;
3. evaluation of diallel crosses among the subgroups together with the parents in replicated field trials; and
4. selection of the most promising cross combinations as potential heterotic patterns using the identification criteria.

A preliminary classification of germplasm into subgroups of genetically similar lines could be based on the geographic origin, morphological data, pedigree information, and breeding history of the crop. An extremely powerful tool for grouping of germplasm are molecular markers such as RFLPs, AFLPs, and

RAPDs (Melchinger, 1993). Using genetic similarities determined from molecular data, relationships between genotypes can be represented graphically by cluster analysis or principal coordinate analysis as demonstrated in Fig. 3–4.

Case 2: Established Heterotic Patterns Are Available

If established heterotic patterns are available, we recommend using selected elite genotypes from them as testers for the production and evaluation of the germplasm to be classified. Based on the testcross performance, populations or lines having similar combining ability and heterotic response could be merged to constitute a new independent heterotic group, if they behave differently from the

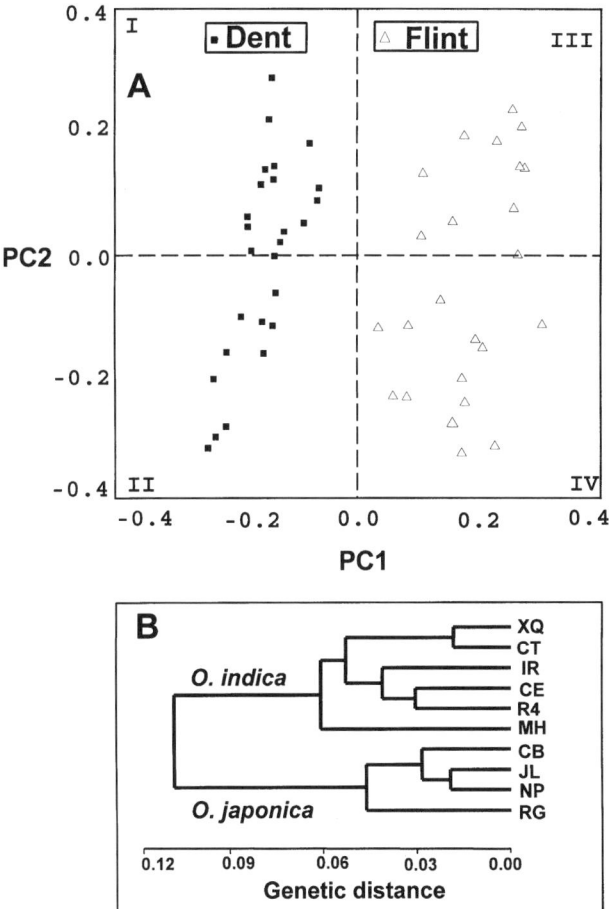

Fig. 3–4. Grouping of germplasm in maize and rice. (A) Principal coordinate analysis of genetic similarities determined from RFLPs in European maize (Messmer et al., 1992): principle coordinate 1 (PC1) separates the Dent lines (squares in Quadrants I and II) from the Flint lines (triangles in Quadrants III and IV); (B) Cluster analysis of genetic distances determined from RAPDs in rice (Xiao et al., 1996): XQ, CT, IR, CE, R4, and MH represent lines from *O. indica*, and CB, JL, NP, and RG represent lines from *O. japonica*.

existing heterotic groups; however, if their behavior is similar to an existing heterotic group, they could be merged with it to enlarge its genetic base. The genetic diversity of lines or populations selected either for constituting a new heterotic group or broadening existing ones should be confirmed by molecular analysis.

For instance, Kauffmann et al. (1982) broadened the genetic base of the newly identified maize heterotic groups Midland and Leaming by using them as testers in crosses with a large number of cultivars. Lines showing a heterotic response similar to Midland in crosses with Leaming were grouped into the Midland heterotic group and vice versa.

CONCLUSIONS

Heterotic groups are the backbone of successful hybrid breeding. The decision regarding the choice of heterotic groups is of fundamental importance and must be made at the beginning of a crop improvement program. Once the heterotic groups are established and improved over a number of selection cycles, it is extremely difficult to develop new competitive ones, because at an advanced stage, the gap in performance between improved breeding materials and unimproved source materials is usually too large. Only if the breeding goals change, is there a good chance to change or develop new heterotic groups with reasonable effectiveness. Moreover, if hybrid breeding is based on CMS, the selection of a heterotic group representing the seed parents (male steriles) and pollen parents (restorers) must be made at an early stage, because considerable efforts are needed for the transfer of CMS and restorer genes. It is evident from the review of various studies that genetic diversity caused by isolation in space and/or time is essential for the establishment of promising heterotic patterns. Heterotic groups should not be considered as closed populations, but should be broadened continuously by introgressing unique germplasm to warrant medium and long-term gains from selection. While characterization of germplasm is possible with molecular markers, heterosis and hybrid performance among unrelated germplasm is generally not predictable and requires field evaluation of crosses among them (Melchinger, 1993).

ACKNOWLEDGMENTS

We are grateful to H.H. Geiger and T. Miedaner for fruitful discussions in preparing this review. The financial support from the Vater und Sohn Eiselen Stiftung, Ulm, to R.K. Gumber is greatly appreciated.

REFERENCES

Adam, D., V. Simonsen, and V. Loeschke. 1987. Allozyme variation in rye, *Secale cereale* L: 2. Commercial varieties. Theor. Appl. Genet. 74:560–565.

Becker, H.C., G.M. Engqvist, and B. Karlsson. 1995. Comparison of rapeseed cultivars and resynthesized lines based on allozyme and RFLP markers. Theor. Appl. Genet. 91:62–67.

Brandle, J.E., and P.B.E. McVetty. 1990. Geographical diversity, parental selection and heterosis in oilseed rape. Can. J. Plant Sci. 70:935–940.

Dhillon, B.S., J. Boppenmaier, W.G. Pollmer, R.G. Herrmann, and A.E. Melchinger. 1993. Relationship of restriction fragment length polymorphisms among European maize inbreds with ear dry matter yield of their hybrids. Maydica 38:245–248.

Dhillon, B.S., P.A. Gurrath, E. Zimmer, M. Wernke, W.G. Pollmer, and D. Klein. 1990. Analysis of diallel crosses of maize for variation and covariation in agronomic traits at silage and grain harvest. Maydica 35:297–302.

Diers, B.W., and T.C. Osborn. 1994. Genetic diversity of oilseed *Brassica napus* germplasm based on restriction fragment length polymorphisms. Theor. Appl. Genet. 88:662–668.

Dudley, J.W., M.A. Saghai-Maroof, and G.K. Rufener. 1991. Molecular markers and grouping of parents in a maize breeding program. Crop Sci. 31:718–723.

Eberhart, S.A., W. Salhuana, R. Sevilla, and S. Taba. 1995. Principles of tropical maize breeding. Maydica 40:339–355.

Engqvist, G.M., and H.C. Becker. 1994. What can resynthesized *Brassica napus* offer to plant breeding. Sver Utsädesförenings Tidsk 104:87–92.

Geiger, H.H., and F.W. Schnell. 1970. Cytoplasmic male sterility in rye (*Secale cereale* L.). Crop Sci. 10:590–593.

Geiger, H.H., and F.W. Schnell. 1975. Experimental basis for breeding hybrid varieties in rye. Hodowla Rosl. Aklim. Nasienn. 19:381–385.

Geiger, H.H., and G. Wahle. 1978. Struktur der Heterosis von Komplexmerkmalen bei Winterroggen-Einfachhybriden. Z. Pflanzenzüchtg 80:198–210.

Goodman, M.M. 1985. Exotic maize germplasm: Status, prospects and remedies. Iowa State J. Res. 59:494–527.

Grant, I., and W.D. Beversdorf. 1985. Heterosis and combining ability estimates in spring oilseed rape (*Brassica napus* L.). Can. J. Genet. Cytol. 27:472–478.

Hallauer, A.R., W.A. Russell, and K.R. Lamkey. 1988. Corn breeding. p. 463–564. *In* G.F. Sprague, and J.W. Dudley (ed.) Corn and corn improvement. 3rd ed. Agron. Monogr. 18. ASA, CSSA, and SSSA, Madison, WI.

Hepting, L. 1978. Analyse eines 7 × 7 - Sortendiallels zur Ermittlung geeigneten Ausgansmaterials für die Hybridzüchtung bei Roggen. Z. Pflanzenzüchtg. 80:188–197.

Jones, J.W. 1926. Hybrid vigour in rice. J. Am. Soc. Agron. 18:423–428.

Kauffmann, K.D., C.W. Crum, and M.F. Lindsey. 1982. Exotic germplasm in a corn breeding program. III. Corn Breeder's School 18:6–39.

Kittlitz, E. von. 1986. Some observations in reciprocal crosses between *Vicia faba major* and *Vicia faba minor*. Biol. Zentralbl. 105:147–153.

Kittlitz, E. von, K.I.M. Ibrahim, P. Ruckenbauer, and L.D. Robertson. 1993. Analysis and use of interpool crosses (Mediterranean × Central European) in faba beans (*Vicia faba* L.): I. Performance of Mediterranean and Central European faba beans in Syria and Germany. Plant Breed. 110:307–314.

Krahling, K. 1987. Utilization of genetic variability of resynthesized rapeseed. Plant Breed. 99:209–217.

Lefort-Buson, M., and Y. Dattee. 1982. Genetic study of some agronomic characters in winter oilseed rape (*Brassica napus* L.): I. Heterosis II. Genetic parameters. Agronomie 2:315–332.

Lefort-Buson, M., B. Guillot-Lemoine, and Y. Dattee. 1987. Heterosis and genetic distance in rape seed (*Brassica napus* L.): Crosses between European and Asiatic selfed lines. Genome 29:413–418.

Lin, S.C., and L.P. Yuan. 1980. Hybrid rice breeding in China. p. 35–51. *In* Innovative approaches to rice breeding. IRRI, Manila, Philippines.

Link, W., C. Dixkens, M. Singh, M. Schwall, and A.E. Melchinger. 1995. Genetic diversity in European and Mediterranean faba bean germplasm revealed by RAPD markers. Theor. Appl. Genet. 90:27–32.

Link, W, W. Ederer, R.K. Gumber, and A.E. Melchinger. 1997. Detection and characterization of two new CMS systems in faba bean. Plant Breed. 116:158–162.

Link, W., B. Schill, A.C. Barbera, J.I. Cubero, A. Filippetti, L. Stringi, E.V. Kittlitz, and A.E. Melchinger. 1996. Comparison of intra- and inter-pool crosses in fababean (*Vicia faba* L.): I. Hybrid performance and heterosis of crosses in Mediterranean and German environments. Plant Breed. 115:352–360.

Martin, J.M., L.E. Talbert, S.P. Lanning, and N.K. Blake. 1995. Hybrid performance in wheat as related to parental diversity. Crop Sci. 35:104–108.

Melchinger, A.E. 1993. Use of RFLP markers for analysis of genetic relationship among breeding materials and prediction of hybrid performance. p. 621–628. *In* International Crop Science I. CSSA, Madison, WI.

Melchinger, A.E., H.H. Geiger, and F.W. Schnell. 1986. Epistasis in maize (*Zea mays* L.): 2. Genetic effects in crosses among early flint and dent inbred lines determined by three methods. Theor. Appl. Genet. 72:231–239.

Melchinger, A.E., H.H. Geiger, G. Seitz, and G.A. Schmidt. 1987. Optimum prediction of three-way crosses from single crosses in maize (*Zea mays* L.). Theor. Appl. Genet. 74:339–345.

Messmer, M.M., A.E. Melchinger, J. Boppenmaier, E. Brunklaus-Jung, and R.G. Herrmann. 1992. Relationship among early European maize inbreds: I. Genetic diversity among flint and dent lines revealed by RFLPs. Crop Sci. 32:1301–1309.

Moll, R.H., J.H. Longquist, J.V. Fortuna, and E.C. Johnson. 1965. The relation of heterosis and genetic divergence in maize. Genetics 52:139–144.

Moll, R.H., W.S. Salhuana, and H.F. Robinson. 1962. Heterosis and genetic diversity in variety crosses of maize. Crop Sci. 2:197–198.

Mungoma, C., and L.M. Pollak. 1988. Heterotic patterns among ten Corn Belt and exotic maize populations. Crop Sci. 28:500–504.

Ogura, H. 1968. Studies on the male sterility in Japanese radish with special reference to the utilization of the sterility towards the practical raising of hybrid seed. Mem. Fac. Agric. Kagoshima Univ. 6:39–48.

Pandey, S., and C.O. Gardner. 1992. Recurrent selection for population, variety and hybrid improvement in tropical maize. Adv. Agron. 48:2–79.

Richey, F.D. 1922. The experimental basis for the present status of corn breeding. J. Am. Soc. Agron. 14:1–17.

Saghai Maroof, M.A., G.P. Yang, Qifa Zhang, and K.A. Gravois. 1997. Correlation between molecular marker distance and hybrid performance in U.S. southern long grain rice. Crop Sci. 37:145–150.

Schnell, F.W. 1978. Progress and problems in utilizing quantitative variability in plant breeding. Plant Res. Dev. 7:32–43.

Schnell, F.W. 1982. A synoptic study of the methods and categories of plant breeding. Z. Pflanzenzüchtg. 89:1–18.

Schnell, F.W. 1992. Maiszüchtung und die Züchtung forschung in der Bundesrepublik Deutschland. Vortr. Pflanzenzüchtung 22:27–44.

Stelling, D., E. Ebmeyer, and W. Link. 1994. Yield stability in faba bean, *Vicia faba* L: 2. Effects of heterozygosity and heterogeneity. Plant Breed. 112:30–39.

Vasal, S.K., G. Srinivasan, F. Gonzalez C., D.L. Beck, and J. Crossa. 1993. Heterosis and combining ability of CIMMYTs quality protein maize germplasm: II. Subtropical. Crop Sci. 33:51–57.

Virmani, S.S., R.C. Aquino, and G.S. Khush. 1982. Heterosis breeding in rice (*Oryza sativa* L.) Theor. Appl. Genet. 63:373–380.

Wallace, H.A., and W.L. Brown. 1956. Corn and its early fathers. Michigan State Univ. Press, East Lansing.

Williams, M.E., 1995. Genetic engineering for pollination control. Tibtech 13:344–349.

Xiao, J., J. Li, L. Yuan, S.R. McCouch, and S.D. Tanksley. 1996. Genetic diversity and its relationship to hybrid performance and heterosis in rice as revealed by PCR-based markers. Theor. Appl. Genet. 92:637–643.

Xiao, J., J. Li, L. Yuan, and S.D. Tanksley. 1995. Dominance is the major genetic basis of heterosis in rice as revealed by QTL analysis using molecular markers. Genetics 140:745–754.

Yuan, L.P. 1992. Development and prospects of hybrid rice breeding. p. 97–105. *In* C.B. You and Z.L. Chen (ed.) Agricultural biotechnology: Proc. Asian Pacific Conf. on Agricultural Biotechnology, China Agriculture Press, Beijing.

Zhang, Q., Y.J. Gao, M.A. Saghai Maroof, S.H. Yang, and J.X. Li. 1995. Molecular divergence and hybrid performance in rice. Molec. Breed. 1:133–142.

Zhang, Q., M.A. Saghai Maroof, T.Y. Lu, and B.Z. Shen. 1992. Genetic diversity and differentiation of *indica* and *japonica* rice detected by RFLP analysis. Theor. Appl. Genet. 83:495–499.

4 Hybrids in Horticultural Crops

Jules Janick
Department of Horticulture
Purdue University
West Lafayette, Indiana

ABSTRACT

Production of hybrids of seed-propagated horticultural crops is one of the most successful plant breeding techniques because it exploits heterosis, promotes homogeneity, and is a way for commercial breeders to control their products without the necessity for legal patents or plant breeder's rights. Various techniques to produce hybrids have been developed depending on the crop including hand emasculation, roguing of staminate plants in dioecious lines, use of gynoecious or highly female lines, cytoplasmic male sterility and genetic male sterility, protogyny, or self incompatibility. Vegetable crops in which 50% or more of the commercial seed crop in the USA are F_1 hybrids includes asparagus (*Asparagus officinalis* L.), broccoli (*Brassica oleracea* L.), Brussels sprouts (*B. oleracea* L.), cabbage (*B. oleracea* L.), carrot [*Daucus carota* L.], collards (*B. oleracea* L.), cucumber (*Cucumis sativus* L.), eggplant (*Solanum melangena* L.), muskmelon (*C. melo* L.), onion (*Allium cepa* L.), pumpkin (*Cucurbita maxima* Duchesne), spinach (*Spinacia oleracea* L.), summer squash (*Cucurbita* sp.), sweet corn (*Zea mays* L.) and popcorn (*Z. mays* L.), tomato [*Lycopersicon esculentum* Mill.], and watermelon [*Citrullus lanatus* (Thunb.) Matsum & Nakai]. Hybrids also dominate the market in a number of ornamental bedding plants and cut flowers especially ageratum (*Ageratum houstonianum* Mill.), carnation (*Dianthus caryophyllus* L.), seed geranium (*Pelargonium hortorum* L.H.Bailey), impatiens (*Impatiens wallerana* Hook.f.), African marigold (*Tagetes erecta* L.), nicotiana (*Nicotiana* L. hybrids), pansy (*Viola wittrockiana* Gams.), petunia (*Petunia hybrida* Hort.Vilm.-Andr.), primula (*Primula vulgaris* Huds.), snapdragon (*Antirrhinum magus* L.), and zinnia (*Zinnia elegans* Jacq.).

General references to heterosis in seed-propagated horticultural crops has been the subject of only a few reviews (Bresnev, 1963; Gabelman, 1974; Pearson, 1986; Reimann-Philipp, 1986) and statistics on hybrid seed production have not been published recently although Pearson (1986) lists data from the USDA Crop Reporting Board for hybrid seed production, 1979–1980 (Table 4–1). In order to determine the status of hybrids in vegetables and ornamentals a survey was made through members of the American Society for Horticultural Science Vegetable Breeding Committee who have responsibility for individual crops (see acknowledgment). A questionnaire was used requesting information on the percentage of

---

Copyright © 1998 Crop Science Society of America, 677 S. Segoe Rd., Madison, WI 53711, USA.
Concepts and Breeding of Heterosis in Crop Plants. CSSA Special Publication no. 25.

Table 4–1. Hybrid vegetable seed production, 1979–1980 (after Pearson, 1983).

Kind/Type	Acres 1979	Acres 1980	Production 1979	Production 1980	Average F_1	Pounds acre^{-1} 1979	Pounds acre^{-1} 1980	Price premium
			(000) of lbs		%			
Broccoli								
Hybrid	344	514	94	170	62	273	330	7
Standard	73	102	87	77		1 191	754	
Cabbage								
Hybrid	366	874	142	282	27	387	322	6.5
Standard	369	793	400	748		1 084	943	
Cauliflower								
Hybrid	36	37	2	12	4	55	324	4
Standard	269	450	128	210		475	466	
Carrot								
Hybrid	396	1 173	50	275	5	126	234	4
Standard	396	7 804	163	5 909		411	757	
Cucumber								
Hybrid	3 009	2 350	1 231	794	41	409	337	6
Standard	3 678	2 321	1 800	1 071		489	461	
Muskmelon								
Hybrid	4	5	0.8	1	--	200	200	20
Standard	3664	2 329	1 280	785		350	337	
Pumpkin								
Hybrid	34	4	15	1.5	5	441	375	6
Standard	363	317	178	138		490	435	
Spinach								
Hybrid	1968	1 367	4 106	2 518	81	2 086	1 841	0.15
Standard	633	344	1 007	494		1590	1436	
Summer Squash								
Hybrid	1 552	1 427	894	730	58	575	511	2.5
Standard	954	839	637	531		667	632	
Winter Squash								
Hybrid	50	53	13	21	7	260	252	3
Standard	518	457	231	180		446	394	
Watermelon								
Hybrid	20	100	0.2	0.5	--	10	5	15
Standard	11 272	10 279	2 892	2 641		256	256	

commercially produced F_1 and F_2 seed, the main systems for producing hybrids, the most important producers of hybrid seed, and a list of any recent citations. Members of Vegetable & Flower Seed Division Executive Committee of the American Seed Association also were queried and contacts were made with Seminis Vegetable Seeds, (Asgrow–Petoseed–Royal Sluis; W. Allen Stevens), Goldsmith Seeds (Niles Riese and Tilly Holtrop), and PanAmerican Seed Co. (Brian E. Corr) with the latter two companies providing information on flower seeds. Thus, while the information presented in this chapter is based on estimates, they are from informed persons. Note that fruit crops, which are vegetatively propagated, are not covered in this chapter; although some rootstocks are seed propagated and F_1 hybrids have been proposed, none are available.

HYBRIDS IN HORTICULTURAL CROPS

Table 4–2. The North American vegetable seed market, 1996; Source: W. Allen Stevens, Asgrow/Petoseed/Royal Sluis.

Crop	Total area	Total seed	Hybrid	Method (currently used mating system)
	acre	lbs	%	
Beets	31 869	341 767	20	Cytoplasmic male sterility
Broccoli	233 365	122 580	100	Self incompatibility
Brussels Sprouts	9 277	4 129	100	Self incompatibility
Cabbage	203 960	99 285	84	Self incompatibility
Carrot	193 802	570 419	56	Cytoplasmic male sterility
Cauliflower	87 979	27 166	33	Self incompatibility
Celery	46 096	12 713	0	
Cucumber, Pickling	113 937	341 811	100	Monoecious and gynoecious
Cucumber, Slicer	107 787	285 290	76	Monoecious and gynoecious
Eggplant	11 606	4 340	50	Hand emasculation
Lettuce	323 943	121 570	0	
Muskmelon, Eastern	32 249	4 855	100	Monoecious, gynoecious and andromonoecious
Muskmelon, Western	178 956	172 236	80	Monoecious, gynoecious and andro monoecious
Onion	253 582	722 151	56	Cytoplasmic male sterility
Pepper, Hot	213 941	465 194	2	Hand emasculation, genetic and cytoplasmic male sterility
Pepper, Sweet	117 150	59 375	36	Hand emasculation
Pumpkin	51 818	105 112	42	Monoecious
Radish	61 410	566 450	0	
Spinach	22 004	521 547	100	Monoecious
Squash, Summer	142 407	477 572	52	Monoecious
Squash, Winter	24 594	95 906	35	Monoecious
Tomato, Fresh market	212 379	16 830	56	Hand emasculation
Tomato, Processing	410 600	191 800	74	Hand emasculation
Watermelon	313 200	224 712	33	Monoecious
Total	3 397 911	5 554 810		

VEGETABLES

A partial list of the North American vegetable seed market is presented in Table 4–2. The percentage of F_1 hybrids varies from zero to 100%. The results of individual crops is presented by family below. Except where stated otherwise, estimates of hybrid seed production here represent U.S. production.

Apiaceae (Umbelliferae)

Carrot (*Daucus carota* L.)

Approximately 90% of fresh market cultivars and 40 to 60% of the canning and freezing types grown in Europe and the USA are F_1 hybrids with about 20 to 40% hybrids for both types in the rest of the world. All hybrid seed is produced by cytoplasmic male sterility. Seed companies that are important producers of hybrid seed include Seminis (Asgrow–Petoseed); Sunseeds Co.; Alf Christianson, Rogers Seed Co., Crookham Company, and Clause (Harris-Moran). Most hybrid seed for Europe and Asian Seed Companies is produced by U.S. growers in the Pacific Northwest (Source: P.W. Simon, Madison, WI).

Celery (*Apium graveolens* L.)

Less than 1% of cultivars are hybrid. Systems of hybrid production is proprietary but is probably cytoplasmic male sterility. Seed companies producing hybrid seed include Takii & Company and Tozer Seed (England; Source: C. Quiros, University of California, Davis).

Parsnip (*Pastinaca sativa* L.)

All are open-pollinated cultivars (Source: P.W. Simon, Madison WI).

Asteraceae

Lettuce (*Lactuca sativa* L.)

Despite the fact that heterosis exists in lettuce, commercial hybrids do not exist because of the difficulty in making crosses. The style elongates as the anther shed pollen on their interior surfaces and pollen must be removed before it can germinate and replaced with pollen from the male parents. Some procedures have been developed by which the proportion of hybrid to selfed seed can be maximized. Genetic male sterility exits in lettuce but no F_1 hybrids have been released from programs underway in Japan. F_2 hybrids have been released from a California seed company but these were highly variable, were unsuccessful, and were later inbred and renamed (Source: E.J. Ryder, USDA-ARS, Salinas, CA).

Brassicaceae

Broccoli (*Brassica oleracea* L., Italica group)

Nearly 100% of commercial seed is hybrid. Most is produced by self incompatibility but a few hybrids produced by cytoplasmic male sterility. Seed producers include Seminis (Asgrow–Petoseed), Rogers Seed Company., Clause (Ferry-Morse–Harris-Moran), Sakata Seed America, American Takii, Shamrock Seed Company., Reed's Seed, and Alf Christianson (Source: M.W. Farnham, USDA-ARS, Charleston, SC).

Brussels Sprouts (*Brassica oleracea* L. Gemnifera group)

About 95% of commercially produced seed is F_1 hybrids (see cabbage for details). The most important producers of hybrid seed are Bejo, Nickerson Zwaan, Seminis (Royal Sluis), Sakata, Sluis and Broot, and American Takii (Source: N.E. De Vos, Tanimura and Antle, Salinas, CA).

Cabbage (*Brassica oleracea* L., Capitata group)

With the exception of a few old cultivars grown in home gardens, all of the cabbage is hybrid. All *Brassica* hybrids are produced by self incompatibility but the change over to male sterility is imminent, probably by 1997. Part of the holdup is due to patent problems with Sandoz involving transformation of the Ogura type sterility to overcome cold temperature sensitivity. Another male sterility type termed "Anand" is under development. The major producers are American Takii, Sakata Seed American, a number of Korean companies, Bejo, Alf Christianson, Seminis (Asgrow–Petoseed), and Clause (Ferry Morse; Source: M.H. Dickson, Cornell University, Geneva, NY).

Cauliflower (*Brassica oleracea* L., Botrytis group)

About 35% of cultivars are F_1 hybrids although the number is rapidly increasing (see cabbage for details; Source: M.H. Dickson, Cornell University, Geneva, NY; and M.W. Farnham, USDA-ARS, Charleston, SC).

Chinese Cabbage (*Brassica campestris* L., Pekinensis group)

Most Chinese cabbage is hybrid (Source: M.H. Dickson, Cornell University, Geneva, NY).

Collards (*Brassica oleracea* L., Acephala group)

About 50 to 75% of seed are F_1 hybrids (Source: M.W. Farnham, USDA-ARS, Charleston, SC).

Kale (*Brassica oleracea* L., Acephala group)

About 50 to 75% of seed is F_1 hybrid (Source: M.W. Farnham, USDA-ARS, Charleston, SC).

Mustard [*Brassica nigra* (L.) Koch]

About 20% of mustard seed is now F_1 hybrids (Source: American Takii).

Radish (*Raphanus sativus* L.)

Only 5% the round red radish seed commonly grown in the USA are F_1 hybrids but they contribute to 100% of greenhouse commercial production in Northern Europe. Hybrid seed is produced by self incompatibility but a shift toward cytoplasmic male sterility is occurring. Producers include Alf Christianson, Rogers Seed Company, and Clause (Harris-Moran), Dorsing Seeds, Agricultural Alternatives, and Flannigan-Kappa. Daikon radish hybrids are mostly open pollinated cultivars but at least five large companies are producing hybrids (Source: J. Behling, Alf. Christianson Seed Company, Lynnwood, WA).

Rutabaga (*Brassica napus* L. Napobrassica group)

No hybrids available (Source: I.L. Goldman, University of Wisconsin, Madison).

Turnip (*Brassica rapa* L.)

More than 70% of seed are F_1 hybrid and are produced by self incompatibility. Sakata Seed America and American Takii produce most of the hybrid turnip (Source: I.L. Goldman, University of Wisconsin, Madison).

Chenopodiaceae

Beet (*Beta vulgaris* L.)

F_1 hybrid seeds make up about 70% of the processing market and 50% of fresh market. Hybrids are produced by cytoplasmic male sterility. Seed producers include Alf Christianson and Seminis (Asgrow–Petoseed; Source: I.L. Goldman, University of Wisconsin, Madison).

Chard (*Beta vulgaris* L., Chicla group)

No hybrids are being sold but breeders at Alf Christianson are evaluating a promising hybrid (Source: I.L. Goldman, University of Wisconsin, Madison).

Spinach (*Spinacia oleracea* L.)

More than 90% of seed is F_1 hybrids. Originally hybrids were produced by roguing staminate plants in dioecious types but now most hybrids are produced from monoecious lines (predominantly pistillate but maintained from late season staminate flowers). Main seed producers are Alf Christianson and Seminis (Asgrow–Petoseed; Source: T.E. Morelock, University of Arkansas, Fayetteville).

Convolvulaceae

Sweet potato [*Ipomoea batatas* (L.) Lam]

All sweet potato is produced clonally (Source: D.R. La Bonte, Louisiana State University, Baton Rouge).

Cucurbitaceae

Cucumber (*Cucumis sativus* L.)

Hybrids are produced in 95% of pickling cucumber, 80% of slicing cucumber, and 60% of garden cucumber. Hybrids are produced by crossing a gynoecious or monoecious inbred with a monoecious inbred, using bee pollination with 0.8 to 1.6 km isolation. Important producers of hybrid seed include Seminis (Asgrow–Petoseed), Sunseeds Company, and Limagrain (Ferry-Morse–Harris-Moran; Source T. Wehner, North Carolina State University, Raleigh).

Muskmelon (*Cucumis melo* L.)

Although standard open-pollinated cultivars are still in catalogues about 80% (eastern USA) to 100% (western USA) of the industry are F_1 hybrids and all current USA efforts on cultivar development are focused on hybrids. Manual emasculation is the main means for producing F_1 hybrids but genetic male sterility and gynoecy are used on a small scale. Seminis (Asgrow–Petoseed–Royal Sluis) and Clause (Ferry-Morse–Harris-Moran) are the major seed producers (Source: J. McCreight, USDA-ARS, Salinas, CA).

Pumpkins and Squash (*Cucurbita pepo* L., *C. moshata* Duchesne ex Poir)

About 50 to 70% of *Cucurbita* seed sold in the USA is F_1 hybrids. In the case of summer squash, about 80% of seed sold in the USA and 30 to 50% in other parts of the world is F_1 hybrid seed. Hybrids are produced with hand emasculated and growth regulators (ethephon) may be used to regulate sex expression (Source: L. Beaver, University of Puerto Rico).

Watermelon [*Citrullus lanatus* (Thunb.) Matsun & Nakai]

About 80% of watermelons are F_1 hybrids in the USA. They are produced by hand pollination of protected pistillate flowers. Seedless watermelons are produced by crossing seed parent tetraploids × diploids. Important producers of hybrid seed are Seminis (Asgrow–Petoseed), American Sunmelon, Rogers Seed Company, and Sunseeds (Source: G.W. Elmstrom, Sunseeds Company, Acampo, CA).

Fabaceae (Leguminosae)

Bean (*Phaseolus vulgaris* L.)

Despite the evidence for heterosis no hybrid seed are available because of the technical difficulty of making crosses. Cytoplasmic genic male sterility, genic control of indehiscent anthers (pollen viable), and associated male and female sterility have been reported (Source: D.P. Coyne, University of Nebraska, Lincoln; and J. Nienhuis, University of Wisconsin, Madison).

Lima Bean (*Phaseolus lunatus* L.)

There are no commercially produced F_1 hybrids of lima bean.

Pea (*Pisum sativum* L.)

There are no commercially produced F_1 hybrids of garden pea despite evidence for the presence of heterosis for yield and other traits. Genetic male sterility exists but not cytoplasmic male sterility (Source: E. Gritton, University of Wisconsin, Madison).

Southernpea [*Vigna unguiculata* (L.) Walp.]

No hybrids are being sold although several studies show that cowpea hybrids can exhibit considerable heterosis for many traits (Source: R.L. Fery, USDA-ARS, Charleston, SC).

Liliaceae

Asparagus (*Asparagus officinalis* L.)

Approximately 50% of the asparagus seed market is F_1 hybrid produced by crossing clonally propagated staminate plants and clonally propagated pistillate plants and 40% is F_2 seed from UC-157. Parent plants are not inbred but the New Jersey staminate parent is inbred one generation to produce *YY* supermales so that all hybrids are staminate. Major hybrid seed producers are the California Asparagus Seed and Transplant Company and Jersey Asparagus Farms (Sources: S.S. Garrison, Rutgers Research Development Center, Bridgeton, NJ; K.C. Sink, and S. Walker).

Garlic (*Allium sativum* L.)

All garlic is produced from clonal cultivars but seed production is now possible (Source P.W. Simon, Madison, WI).

Leek (*Allium ameloprasum* L.)

About 20% of seed are estimated to be F_1 hybrid, presumably from cytoplasmic male sterility (Source: P.W. Simon, Madison, WI).

Onion (*Allium cepa* L.)

About 65% of seed sold is F_1 hybrids and all are produced by cytoplasmic male sterility. The 1995 worldwide commercial value of hybrid onion seed was estimated to be $148 million. Southern production regions in the USA still have significant production of open-pollinated cultivars primarily because they are in sterile cytoplasm. Northern production areas are at least 90% hybrid. Principal seed producers are Seminis (Asgrow–Petoseed), Rio Colorado Seeds, and Sunseeds Company (Source: M.J. Havey, USDA-ARS, Madison, WI).

Shallots (*Allium cepa* L. Aggregatum group)

Most seed is open-pollinated but a single hybrid is now available from a Dutch company (Source: M.J. Havey, USDA-ARS, Madison, WI; and P.W. Simon, Madison, WI).

Malvaceae

Okra (*Hibiscus esculentus* L.)

Only about 5 to 10% of domestic USA seed is hybrid and about 2% of world production. Hybrid seed production is by hand emasculation. Seed producers include Seminis (Petoseed), Sandoz Corporation, and various seed companies in India, Japan, and Korea (Source: R.L. Jarret, USDA-ARS, Griffin, GA).

Poaceae (Graminae)

Popcorn (*Zea mays* L.)

More than 99% of popcorn is F_1 hybrid. Seed producers include the Agricultural Alumni Seed Improvement Association, Crookham Company, Iowa Acres, McHone Seed Company, Meade Seed, Schlessman Seed Company (Source: W.F. Tracy, University of Wisconsin, Madison).

Sweet corn (*Zea mays* L.)

About 99% of sweet corn and popcorn are F_1 hybrids. Manual detasseling is the common method of production but about 20% are produced by cytoplasmic male sterility. Important hybrid seed producers include Seminis (Asgrow), Crookham Company, IFS, Clause (Ferry-Morse–Harris-Moran), Rogers Seed Company (Source: W.F. Tracy, University of Wisconsin, Madison).

Solanaceae

Eggplant (*Solanum melongena* L.)

About 60% of eggplant seed is F_1 hybrid and is produced by manual emasculation especially in Asia. Takii is the major developer of hybrid seed (Source: B.L. Pollack, Oceanside, CA).

Pepper (*Capsicum annuum* L.)

About 40% of commercially produced sweet pepper are F_1 hybrids but only about 2% of hot peppers are hybrid. Hybrids are mostly produced by manual emasculation but a few are produced by cytoplasmic male sterility. Important producers of hybrid seed include Abbott & Cobb, Seminis (Asgrow–Petoseed), Rogers Seed Company, Sakata Seed America, and Sunseeds Company (Source: B. Villalon, Texas Agricultural Experiment Station, Weslaco; and M.A. Stevens, Seminis Vegetable Seeds, Woodland, CA).

Potato (*Solanum tuberosum* L.)

Most potatoes are clonally propagated. Potatoes produced by true seed are not produced by crossing inbreds because inbreeding depression is too severe (Source: D. Douches, Michigan State University, East Lansing).

Tomato (*Lycopersicon esculentum* Mill.)

In the USA, effectively 100% of fresh market and 80% of processing tomato are F_1 hybrids. Most hybrids are produced by manual emasculation but a few are assisted with the use of genetic male sterility. Most of the fresh market tomato hybrids imported from Mexico are *rin* heterozygotes to increase shelf life and because of superior genetic background have outperformed the 'FlavrSavr' tomato. F_1 hybrids offer the advantages of complementation for disease resistance and other characters, heterosis for yield, and especially yield stability. Major seed

Table 4–3. Estimation of hybrid seed in ornaments, 1996; source: T. Holtrop and N. Riese, Goldsmith Seeds.

Common name	Scientific name	F_1 hybrid %	Method
African marigold	*Tagetes erecta* L.	90	Genetic male sterility (apetalous)
Ageratum	*Ageratum houstonianum* Mill.	80	Self incompatibility
Begonia	*Begonia* × *semperflorens-cultorum* Hort.	100	Emasculation
Carnation	*Dianthus caryophyllus* L.	80	Genetic male sterility
Cyclamen	*Cyclamen persicum* Mill.	60	Emasculation
Dahlia	*Dahlia cav* hybrids	10	Self incompatibility
Dianthus	*Dianthus* L. hybrids	70	Genetic male sterility
Gazania	*Gazania rigens* (L.) Gaaertn.	50	Self incompatibility
Impatiens	*Impatiens walleriana* Hook.f.	100	Cytoplasmic male sterility
Nicotiana	*Nicotiana* L. hybrids	100	Emasculation
Pansy	*Viola* × *wittrockiana* Gams.	85	Emasculation
Petunia	*Petunia* × *hybrida* Hort. Vilm. Andr.	100	Emasculation (cytoplasmic male sterility linked to nonflowering)
Phlox	*Phlox drummondii* Hook.	10	Emasculation
Portulaca	*Portulaca grandiflora* Hook.	50	Emasculation
Primula	*Primula* × *polyantha* Huds.	80	Self incompatibility
Seed geranium	*Pelargonium* × *hortorum* L.H. Bailey	100	Genetic male sterility
Snapdragon			
cut	*Antirrhinum majus* L.	99	Emasculation
garden	*Antirrhinum majus* L.	80	Emasculation
Zinnia	*Zinnia elegans* Jacq.	75	Genetic male sterility (apetalous)

companies include Seminis (Asgrow–Petoseed-Royal Sluis) and Hazera (Israel; Sources: J. Scott, University of Florida, Bradenton; and M.A. Stevens, Seminis Vegetable Seeds, Woodland, CA).

Floral Crops

A great number of flower crops, particularly those grown as bedding plants are seed propagated. The species in which F_1 hybrids are marketed and the principal method of hybrid production are listed in Table 4–3. In the agricultural sense, F_1 hybrids are crosses between divergent inbreds. In flower breeding, some of the hybrids are produced from lines that are inbred only a few generations, and in many cases are the offspring of parents that do not differ greatly morphologically. Seed parents are usually chosen with the same flower color of the hybrid so that occasional selfs will not look like off types. According to R. J. Griesbach, Floral and Nursery Plant Research, U.S. National Arboretum, some flower breeders and marketers use the term hybrid in the botanical sense as the offspring of dissimilar parents that may differ by a few genes. As a result, a number of so-called F_1 hybrid (such as petunia and impatiens) do now show obvious morphological segregation when selfed; however, consultation with Tilly Holtrop, breeder at Goldsmith Seeds, and Brian Corr of Pan American Seeds, indicate that all of their hybrids are the result of divergent crosses.

SUMMARY AND CONCLUSIONS

1. The production of F_1 hybrids in vegetables and bedding plants is a dynamic part of the seed business and is increasing in both the number of species in which hybrids are sold and the percentage of seed of individual crops. In many crops, almost the entire commercial crop is produced from hybrid seed. The reasons are both biological (heterosis) and commercial (the ability to biologically protect germplasm).
2. Despite the importance of F_1 hybrids, research and breeding on heterosis has quickly moved from the public to the private sector. This has resulted in a dearth of archived publications on both the statistics of hybrid seed production as well as evaluation of hybrids and heterosis. For example, there are practically no papers on the extent of heterosis for most vegetable or floral crops.
3. F_1 hybrids that are important for cross pollinated plants also have found a place for self-pollinated plants such as tomato.
4. Many methods are used to produce hybrids in horticultural crops including emasculation, self-incompatibility, genetic and cytoplasmic male sterility, and the production of gynoecious or highly pistillate monoecious plants.
5. Despite the fact that hybrids have been a part of agriculture since the 1920s, by and large the public is not clear what hybrids are and how they are obtained. This has resulted in many questioning the ethicality of seed producers controlling the distribution of hybrids, concerns over the lack of diversity by moving to hybrids (despite the fact that the use of dissimilar parents is necessary to produce hybrids), and the feeling of some that hybrids are dangerous and even immoral. One of the reasons is that the symbol F_1 (first filial generation) is obscure.

ACKNOWLEDGMENT

I gratefully acknowledge assistance of the following persons who contributed information to this chapter: Linda Beaver, University of Puerto Rico, Mayaguez; Jonathan P. Behling, Alf. Christianson Seed Company, Lynnwood, WA; Brian E. Corr, PanAmerican Seed Company, Chicago, IL; Dermot P. Coyne, University of Nebraska, Lincoln; Neal E. De Vos, Tanimura and Antle, Salinas, CA; Michael H. Dickson, Cornell University, Geneva, NY; David Douches, Michigan State University, East Lansing; Gary W. Elmstrom, Sunseeds Company, Acampo, CA; Mark W. Farnham, USDA-ARS, Charleston, SC; Richard L. Fery, USDA-ARS, Charleston, SC; Stephen A. Garrison, Rutgers Research and Development Center, Bridgeton, NJ; Irwin L. Goldman, University of Wisconsin, Madison; Robert J. Griesbach, U.S. National Arboretum, Beltsville, MD; Earl Gritton, University of Wisconsin, Madison; Michael J. Havey, USDA-ARS, Madison, WI; Tilly Holtrop, Goldsmith Seeds, Gilroy, CA; Robert L. Jarret, USDA-ARS, Griffin, GA; Don R. La Bonte, Louisiana State University, Baton Rouge; James D. McCreight, USDA-ARS, Salinas, CA; Teddy E. Morelock, University of Arkansas, Fayetteville; James Nienhuis, University of Wisconsin, Madison; Bernard L. Pollack, Oceanside, CA; Carlos Quiros,

University of California, Davis; Niles Riese, Goldsmith Seeds, Gilroy, CA; Edward J. Ryder, USDA-ARS, Salinas, CA; John W. Scott, University of Florida, Bradenton; Philip W. Simon, Madison, WI; M. Allen Stevens, Seminis Vegetable Seeds, Woodland, CA; William F. Tracy, University of Wisconsin, Madison; Benigno Villalon, Texas Agricultural Experiment Station, Weslaco; Todd Wehner, North Carolina State University, Raleigh.

REFERENCES

Bresnev, D.D. 1963. Heterosis in vegetable crops. Zuchter 33:134–138.

Gabelman, W.H. 1974. F_1 hybrids in vegetable production. p. 419–428. XIX Int. Hort. Congr., Warsaw.

Pearson, O.H. 1983. Heterosis in vegetable crops. p. 138–188. In R. Frankel (ed.) Heterosis. Monographs on Theoretical and Applied Genetics. Vol 6. Springer-Verlag, New York.

Reimann-Philipp, R. Heterosis in ornamentals. p. 234–259. In R. Frankel (ed.) Heterosis. Monographs on Theoretical and Applied Genetics, Vol 6. Springer-Verlag, New York.

5   Apparent Overdominance in Natural Plant Populations

Jeffry B. Mitton

*Department of Environmental, Population,
and Organismic Biology
University of Colorado
Boulder, Colorado*

ABSTRACT

Surveys of enzyme polymorphisms in plant species provide data to search for evidence of natural selection. Studies of enzyme kinetics typically reveal differences in the performances of the alternate genotypes at a locus, providing an opportunity for natural selection among genotypes within populations. Comparisons among genotypic distributions from successive stages in the life cycle provide sensitive tests for natural selection. Such tests have revealed consistent patterns of overdominance in annual ryegrass (*Lolium multiflorum* Lam.) and slender wild oats (*Avena barbata* Pott ex Link). A survey of selection coefficients has revealed that overdominance is common in plants. Samples of mixed age classes of coniferous forest trees usually have genotypic distributions that fit Hardy-Weinberg expectations, for these trees produce almost all of their seed by outcrossing; however, when samples are restricted to old trees, large trees, or trees producing cones, genotypic distributions typically contain excesses of heterozygotes. Such heterozygote excesses cannot be produced by selection against inbred genotypes, but must be produced by selection favoring heterozygous genotypes. Moreover, comparisons of trees selected for seed orchards with trees in natural populations reveal that foresters consistently choose highly heterozygous trees. Many observations suggest that overdominance will be expressed more frequently or to a greater degree during environmental stress. For example, resistance to various forms of airborne pollution increases with allozyme heterozygosity in conifers. In pinyon pine (*Pinus edulis* Engelm.) trees resistant to herbivorous insects are significantly more heterozygous than susceptible trees.

Inbreeding depression and heterosis have been under investigation for decades (Wright, 1977; Frankel, 1983), but no consensus exists concerning the genetic mechanisms underlying these phenomena. Inbreeding depression, commonly observed when typically outcrossing populations inbreed, is expressed as inferior growth rate, viability, developmental stability, fecundity, and fertility (Lerner, 1954; Crumpacker, 1967; Franklin, 1972; Wright, 1977; Simmons & Crow, 1977; Sorensen & Miles, 1982; Charlesworth & Charlesworth, 1987; Crow & Simmons, 1983; Lande & Schemske, 1985). Heterosis is the luxuriance and supe-

---

Copyright © 1998 Crop Science Society of America, 677 S. Segoe Rd., Madison, WI 53711, USA. *Concepts and Breeding of Heterosis in Crop Plants.* CSSA Special Publication no. 25.

rior performance of the highly heterozygous offspring of crosses between differentiated populations (Shull, 1948; Lerner, 1954; Frankel, 1983).

Two contending hypotheses, the dominance hypothesis and the overdominance hypothesis, are compatible with empirical results from inbreeding depression and heterosis. The dominance hypothesis attributes inbreeding depression to the enhanced expression, during inbreeding, of deleterious recessive alleles. Because completely inbred strains tend to be fixed for deleterious alleles at different loci, crosses between strains produce genotypes bearing deleterious recessive alleles in the heterozygous condition. Heterozygosity masks a majority of the deleterious alleles, producing heterosis. Overdominance is the superior fitness of heterozygotes compared with homozygotes. The overdominance hypothesis attributes inbreeding depression to the loss of heterozygosity, and heterosis to the summation of overdominance over many loci.

Biologists agree that the dominance hypothesis explains a substantial proportion of inbreeding depression and heterosis. Most biologists agree that protein polymorphisms can be neutral at some times, but most doubt that overdominance is either sufficiently common or strong to make the overdominance hypothesis relevant to our understanding of inbreeding depression and heterosis. This review evaluates this commonly-held belief by summarizing examples of overdominance in natural plant populations to assess whether it is sufficiently common in natural plant populations to make a contribution to inbreeding depression and heterosis.

Population geneticists commonly refer to three forms of overdominance: overdominance, marginal overdominance, and associative overdominance. Overdominance is a pattern of relative fitnesses in one or more environments in which the relative fitness of the heterozygote is superior to the relative fitnesses of the homozygotes. Marginal overdominance refers to the fitness advantage accruing to heterozygotes when the relative fitnesses of heterozygotes vary less than homozygotes through time or over space in heterogeneous environments (Wallace 1959, 1981, p. 247). Fluctuations in relative fitnesses can produce marginal overdominance, even when the fitnesses of heterozygotes are intermediate between the fitnesses of the homozygotes in every set of environmental conditions. Associative overdominance (Kimura & Ohta, 1971; Charlesworth, 1991; Hedrick & Savolainen, 1996) is the apparent overdominance at neutral loci that comes from either selected loci in linkage disequilibrium with the marker locus, or fitness variation resulting from variation in inbreeding coefficients within a population. In some circumstances, it is possible to reject the hypothesis that the apparent overdominance is attributable to variation of inbreeding coefficients. But in most population biology studies, it is not possible to distinguish among overdominance, marginal overdominance, and associative overdominance caused by selected loci in linkage disequilibrium with the marker locus. For the purposes of this review, it is not important to distinguish among the three forms of overdominance. The point addressed in this review is whether overdominance is common in plant populations—it is irrelevant whether overdominance occurs in any particular set of environmental conditions or whether it accumulates as environmental conditions fluctuate. Similarly, it is irrelevant whether the selection acts on enzyme polymorphisms or loci in linkage disequilibrium with them. For this reason, the title of this review uses the term *apparent overdominance*, and for simplicity, I refer to the superior fitnesses of heterozygotes as overdominance, regardless of the underlying mechanism of fitness determination.

LOCI UNDER BALANCING SELECTION

A general prediction was reported by Ginzburg (1979, 1983) and Turrelli and Ginzburg (1983) on the relationship between fitness and heterozygosity. For loci whose variation is maintained by balancing selection, average fitness increases with the number of heterozygous loci. The data best suited for testing this prediction are the surveys of enzyme polymorphisms in plant populations (Hamrick & Godt, 1990).

The criteria to identify the impact of natural selection on an enzyme polymorphism were established by Clarke (1975) and Koehn (1978). Firstly, there must be functional differences among the enzymes. Functional differences are identified with analyses of enzyme kinetics (Hall & Koehn, 1983). Secondly, the kinetic differences must produce physiological differences among the genotypes. Finally, the physiological differences must interact with ecological conditions to cause variation in one or more components of fitness.

Evolutionary biologists have analyzed kinetic variation in more than a dozen enzyme, hemoglobin, haptoglobin, and transferrin polymorphisms (Koehn et al., 1983; Frelinger, 1972; Templeton, 1982; Zera et al., 1983; Snyder, 1981; Powers et al., 1994; Mitton, 1997). In the majority of the studies reported, kinetic differences were detected among the segregating genotypes. The majority of enzyme polymorphisms that were analyzed kinetically are in animals. Enzyme kinetics and measures of performance have been reported in ryegrass (*L. multiflorum* and *L. perenne*) and in Norway spruce [*Picea abies* (L.) Karsten]. Although biases in the selection of polymorphisms for kinetic analyses and in the publication of results make it difficult to determine what proportion of enzyme polymorphisms have significant kinetic variation, clearly, some do.

Ryegrasses

Enzyme kinetic variation of 6-phosphogluconate dehydrogenase (*6PGD*) influences flux through metabolic pathways, respiration, and survival during heat stress in perennial ryegrass (*L. perenne*; Rainey et al., 1987, 1990; Rainey-Foreman & Mitton, 1995). Enzyme kinetic studies in perennial ryegrass were inspired by the reports that rates of dark respiration are highly heritable and are inversely correlated with growth rate in perennial ryegrass (Wilson, 1982; Wilson & Jones, 1982; Day et al., 1985). Studies of the regulation of dark respiration suggest that control is not in mitochondria, but in glycolysis (Day et al., 1985). Although *6PGD* does not function in glycolysis, the reaction catalyzed by 6PGD produces ribulose 5-phosphate, which inhibits phosphoglucose isomerase and consequently reduces flux through glycolysis (Noltman, 1972).

In perennial ryegrass, two common alleles, referred to as alleles 1 and 2 for their electrophoretic mobilities, segregate at the *6PGD* locus (Rainey et al.,

Table 5–1. Peroxidase genotypic frequencies and fitnesses estimated from viabilities at two sites at Lake Berryessa, CA.†

	Genotype			Fitness		
	11	12	22	11	12	22
Site A						
Seedlings	257	1355	2954			
Adults	19	171	323	0.91	1.00	0.99
Site B						
Seedlings	278	626	758			
Adults	22	87	76	0.74	1.00	0.85

† From these and other allozyme data, the value of the inbreeding coefficient, F, was estimated to be 0.102. Seedling genotypic frequencies were adjusted, prior to the estimation of the fitnesses, to remove this level of inbreeding.

1987). Kinetic studies measured the Michaelis constant, a measure of the binding affinity of an enzyme for its substrate. Michaelis constants are heterogeneous among genotypes (Rainey-Foreman & Mitton, 1995), and studies with double-labeled C revealed that the fluxes through glycolysis are heterogeneous among the genotypes (Rainey-Foreman & Mitton, 1995). Rates of dark respiration differ among *6PGD* genotypes, with the rate of homozygous *6PGD-11* individuals exceeding the rate of *6PGD-22* homozygotes by >50% ($P < 0.01$; Rainey et. al., 1990). The Q_{10} of dark respiration, or the increase in dark respiration with a 10°C increase in temperature, is highest in *6PGD-11*, intermediate in *6PGD-12*, and lowest in *6PGD-22*. The adaptive significance of the variation in Q_{10} was tested in plants with *12* and *11* genotypes supplied with adequate water and light, but stressed with high day-time temperatures for a week. As predicted by the values of Q_{10}, *12* heterozygotes suffered lower mortality and less leaf damage than the *11* homozygotes.

Although enzyme kinetic studies are limited to 6PGD in perennial ryegrass, demographic studies suggest that several enzyme polymorphisms are associated with viability in annual ryegrass (Mitton, 1989, 1997). Two population samples were taken from hillsides on the eastern shore of Lake Berryessa, in the coastal mountains of northern California. Panicle samples were taken in the field, and open-pollinated families were grown in the greenhouse. The maternal genotypes were inferred from the distribution of genotypes in families. If we assume that these populations were at equilibrium, then the zygotic distributions in the two phases of the life cycle can be used to estimate viability differentials between the seedling and the adult stages. In eight comparisons of seedling and adult genotypic distributions, five were highly significantly different (Mitton, 1997). In each of these comparisons, such as the representative examples in Table 5–1, the heterozygote had the highest fitness.

The adaptive distance model (Smouse, 1986) provides a test to make inferences concerning the presence or absence of selection at polymorphic loci. The adaptive distance test is used to determine whether the pattern of fitnesses estimated for the genotypes is consistent with the allelic frequencies in the field. Such analyses are particularly useful in cases where correlations between heterozygosity and components of fitness cause an investigator to wonder whether selection is acting directly at a locus, or a small group of loci.

Table 5–2. Overdominance and adaptive distances.†

	Genotypes		
	AA	Aa	aa
Frequency	p^2	$2pq$	q^2
Fitness	$-s$	1	$-t$
Adaptive distance	$/p$	0	$1/q$

† An overdominant locus with two alleles, A and a, whose frequencies are p and q, and with selection intensities s and t against the AA and aa genotypes, respectively. The adaptive distance of the heterozygote is defined to be zero, and the adaptive distances of the homozygotes are defined as the inverses of the frequencies of their alleles. When plotted on an adaptive distance plot (see Fig. 5–1) the points for the three genotypes fall on a straight line with a negative slope $\{b = [-st/(s + t)]\}$ equal to the segregational genetic load.

Smouse (1986) defined a new variable, the adaptive distance, in which the heterozygote is assigned a value of zero, and each homozygote is assigned a value that is the reciprocal of the frequency of its allele (Table 5–2). For each genotype, the natural log of relative fitness is plotted on the ordinate, and the adaptive distance is plotted on the abscissa (Fig. 5–1). If the locus is overdominant, allelic frequencies are at equilibrium, and selection is acting directly upon the polymorphism, then the points on the graph define a line. The slope of the line $\{b = [-st/(s + t)]\}$ is proportional to the intensity of selection acting on the locus—the stronger the overdominance, the steeper the slope.

Adaptive distance plots for the peroxidase polymorphism in annual ryegrass are presented in Fig. 5–1, and the original data are presented in Table 5–1. Five enzyme polymorphisms were used to measure the mating system. From all

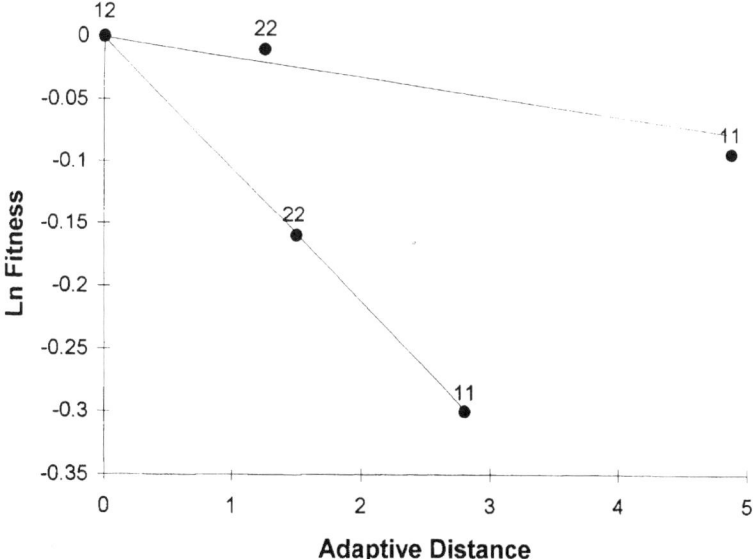

Fig. 5–1. Adaptive distance plots for the peroxidase polymorphism in two population samples of annual ryegrass at Lake Berryessa, CA. 11, 12, and 22 are genotypes at the peroxidase polymorphism (see Table 5–1) defined by their electrophoretic mobilities.

of the genotypic distributions, F, the inbreeding coefficient was estimated to be 0.102. Because selection against inbred genotypes produces adaptive plots that appear to reveal overdominance, the genotypic distributions of seedlings were adjusted to genotypic distributions expected with random outcrossing. After the zygotic distributions had been adjusted to remove inbreeding, the differences between genotypic frequencies in seedlings and adults were used to estimate the viabilities of the three genotypes (Table 5–1). The relative viabilities were used as estimates of fitness, and the allelic frequencies were used to calculate the adaptive distances. The intensities of selection differed between the two sites, but this is not surprising, for ecological geneticists have long recognized that selection varies in both space and time. Both adaptive distance plots are consistent with the hypothesis of overdominance at the peroxidase polymorphism.

Norway Spruce

Deteriorating air quality is challenging the forests of Europe, causing either mortality or declining vigor in several tree species (Scholz et al., 1989; Giannini, 1991). Resistance to air pollution in Norway spruce appears to be influenced by variation at *Pepc*, the locus coding for phosphoenol pyruvate carboxylase (Bergmann & Scholz 1989). PEPC is involved in a CO_2 fixation system, and it plays roles in carbohydrate turnover in guard cells and in the stomatal movements of needles. Initially, demographic studies of various age classes of Norway spruce provided data leading to the hypothesis that PEPC heterozygotes were resistant to air pollution, primarily in the form of SO_2. In natural stands of spruce with low levels of SO_2, genotypic frequencies do not differ between mature spruces and their naturally regenerated seedlings. In sites receiving high loads of SO_2 and other air-borne pollutants in Germany, however, genotypic frequencies in seedlings differed from those in surrounding mature trees, but only at the *Pepc* locus, and not at any of the other three enzyme-coding loci monitored in the study. Two alleles segregate at *Pepc* in the population, and the seedling and adult genotypic frequencies were compared at two polluted sites. At one site, 17% of mature trees were heterozygotes, but 40% of seedlings were heterozygotes, and at the other site, heterozygosities in mature trees and seedlings were 22 and 36%, respectively. Furthermore, the inbreeding coefficients were heterogeneous among the allozyme loci in this study. The other allozyme polymorphisms revealed F_{is} to be approximately 0.10, revealing a small amount of selfing or inbreeding. But F_{is} at PEPC was approximately −0.20, indicating that natural selection had favored heterozygotes at this locus. These data indicate that natural selection differed between relatively clean and polluted sites, and that in polluted sites, selection increased heterozygosity dramatically in the production of seed or the establishment of seedlings, or both.

A subsequent kinetic study of the enzyme products of the *Pepc* locus revealed biochemical differences that might have been the target of selection (Rothe & Bergmann, 1995). Enzyme kinetic studies compared the enzyme products from the three *Pepc* genotypes for the Michaelis constant (K_m), the maximum velocity of the enzyme reaction (V_{max}), and a measure of biochemical efficiency (V_{max}/K_m). When kinetic studies used phosphoenol pyruvate as the substrate, no differences were detected between the homozygotes, but the enzyme from heterozygotes was different from the two homozygotes for K_m, V_{max}, and V_{max}/K_m. The enzyme efficiency, as estimated from V_{max}/K_m, was 60% greater in the heterozygote than it was in the homozygotes. The enhanced enzyme efficiency of heterozygotes may explain the greater resistance to air borne SO_2. In the populations examined, the enhanced heterozygosity in seedlings was interpreted as a consequence of natural selection by recent increases in SO_2 (Rothe & Bergmann, 1995).

Deteriorating air quality is challenging the forests of Europe, causing either mortality or declining vigor in several species (Scholz et al., 1989; Giannini, 1991). While the study above provides unique insights regarding the association between enzyme kinetics and natural selection, other genetic studies which have examined stresses imposed by pollution have reported similar observations. Stressful conditions appear to accentuate differences among genotypes, often producing overdominance or heterosis (Parsons, 1971, 1973, 1987; Mitton & Grant, 1984). Similarly, stress enhances positive correlations between enzyme heterozygosity and components of fitness (Koehn & Shumway, 1982; Diehl, 1988; Scott & Koehn, 1990; Teska et al., 1990). Moreover, comparisons between resistant and susceptible trees from the same stands suggest that heterozygous genotypes confer resistance to air-borne pollutants in Norway spruce (Scholz & Bergmann, 1984; Bergmann & Scholz, 1984, 1987, 1989; Prus-Gowacki & Godzik, 1995), beeches (*Fagus sylvatica* L.; Muller-Stark, 1985), and Scots pine (*P. sylvestris* L.; Geburek et al., 1987).

EXCESS HETEROZYGOSITY IN BIG, OLD, AND SELECTED TREES

When random samples of coniferous forest trees are examined, genotypic proportions generally fit Hardy-Weinberg expectations; however, when only mature, or only the largest, or only the oldest trees are sampled, analyses often reveal excesses of heterozygotes (summarized in Mitton & Jeffers, 1989). Significant excesses of heterozygotes have been reported in mature stands of ponderosa pine (*P. ponderosa* ex P. Lawson & Lawson; Linhart et al., 1981), jack pine (*P. banksiana* Lambert; Cheliak et al., 1985), black spruce [*Picea mariana* (Miller) B.S.P.; Boyle & Morgenstern, 1986; Yeh et al. 1986], Monterey pine (*Pinus radiata* D. Don; Plessas & Strauss, 1987), Douglas-fir [*Pseudotsuga menziesii* (Mirbel) Franco; Shaw & Allard, 1982) balsam fir [*Abies balsamea* (L.) Miller; Neale & Adams, 1985), Polish larch (*Larix decidua* Miller; Lewandowski et al., 1991), Bosnian pine (*Pinus leucodermis* Antoine; Morgante et al., 1993), quaking aspen (*Populus tremuloides* Michaux; Cheliak & Dancik, 1982; Jelinski & Cheliak, 1992; Mitton & Grant, 1995) and the stone pines (*P. sibirica* and *P. cembra* L.; Politov & Krutovskii 1994). As expected, excess heterozygosity has not been detected in all samples of sexually mature conifers; environmental conditions and, consequently, natural selection and its genetic consequences vary in both space and time. For example, Goncharenko et al. found excesses of heterozygosites in dwarf Siberian pine [*P. pumila* (Pall.) Rege; 1993a], but not in

Siberian pine (*P. sibirica* Du Tour; 1993b) or in Scotch pine (*Pinus sylvestris* L.; 1995). If selection acts solely to eliminate selfed genotypes, then all outcrossed genotypes would have equal fitnesses, and selection against inbred genotypes would result in an F of zero. Whereas the decline in values of F from initial positive values to zero is consistent with selection against inbred individuals, the appearance of excesses of heterozygotes in a population must be due to a different process that favors heterozygotes (Shaw & Allard, 1982; Mitton & Jeffers, 1989; Mitton, 1997).

It appears that foresters use the advantageous traits associated with highly heterozygous genotypes when they select trees for breeding. When foresters established orchards of Engelmann spruce (*Picea engelmannii* Parry ex Engelm.; Mitton & Jeffers, 1989), Norway spruce (Bergmann & Ruetz, 1991), Sitka spruce [*Picea sitchensis* (Bong.) Carrière; Chaisurisri & El-Kassaby, 1994] and Douglas-fir (El-Kassaby & Ritland, 1996; Meinartowicz & Lewandowski, 1994) they selected trees based on economically important criteria; however, they unknowingly chose trees with higher individual heterozygosities than would be present in a random sample of trees.

Heterozygosity also may confer an advantage for seed production in trees. An analysis of the mating system of teak (*Tectona grandis* L.f.; Kertadikara & Prat, 1995), indicated the occurrence of a high level of outcrossing, and when the mating system and degrees of genetic variability of teak are considered, the average tree is expected to have a heterozygosity of $H = 0.44$; however, when the subset of trees producing seed were examined, the heterozygosity was significantly higher ($H = 0.64$) than the average ($H = 0.44$).

RESISTANCE TO HERBIVORES

Pinyon pines on the cinder soils around Sunset Crater, near Flagstaff, AZ, experience chronic water and nutrient stress, and also sustain higher densities of herbivores than pines living nearby on the normal sandy loam soils (Whitham & Mopper, 1985; Mopper & Whitham, 1986; Mopper et al., 1991); however, the impact of the stress and herbivory is not uniform among the trees on the cinder soils. Some trees sustain little or no damage from herbivores, while others are trimmed so regularly by the stem moth (*Dioryctria albovitella*) that they assume a different growth form (Whitham & Mopper, 1985; Mopper & Whitham, 1986). By defoliating the tips of branches, the stem moths increase internal branching, producing an atypical, densely packed, closely trimmed growth form that records chronic herbivory. Resistant trees were significantly more heterozygous than susceptible trees at two of four allozyme loci (Mopper et al., 1991; Fig. 5–2). In addition, in both susceptible and resistant trees, older trees were more heterozygous than younger trees, revealing viability differentials favoring heterozygotes.

A SUMMARY OF FITNESSES IN PLANT POPULATIONS

A survey of research in plant population biology revealed that fitness differentials in natural populations of plants generally favor enzyme heterozygotes

(Lesica & Allendorf, 1992). Thirty-eight selection coefficients estimated for allozyme loci were gleaned from eight studies of plants. Twelve of 38 selection coefficients were significantly different from zero, and all revealed higher fitnesses in heterozygotes. When the relative fitness of heterozygotes was set at 1.0, the average fitness of homozygotes at the 38 loci was 0.77. While this intensity of selection is considered to be high by most population geneticists, it is midrange for morphological studies of natural populations (Endler, 1986).

The genes included in this review of the literature code for enzymes, which catalyze metabolic reactions; however, reports of overdominance are not limited to enzyme loci. A locus with strong overdominance for viability in rapid cycling brassica [*Arabidopsis thalliana* (L.) Heynh.] was detected by interval-mapping using molecular markers (Mitchell-Olds, 1995). Homozygotes at this locus have viabilities 50% lower than heterozygotes. Although the locus was mapped to a short segment on chromosome I, the study did not elucidate the gene product or its function.

From the general pattern of overdominance identified in plant populations by Lesica and Allendorf (1992), and by the number of species of trees exhibiting

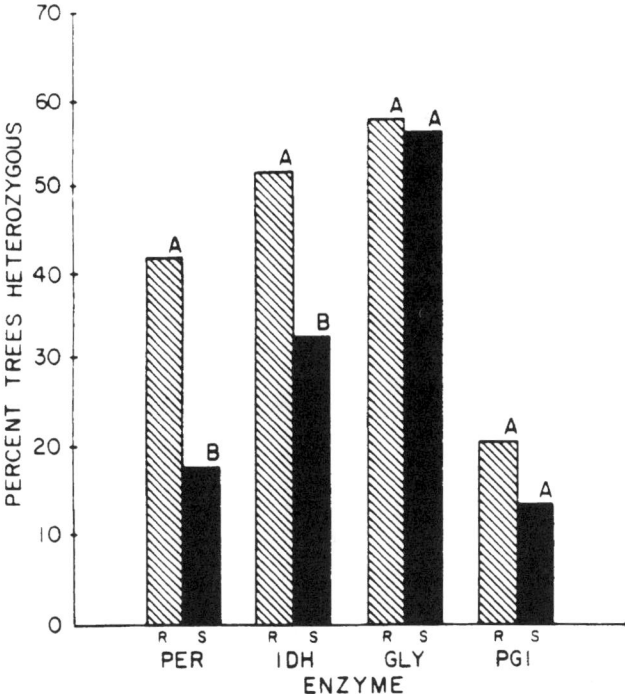

Fig. 5–2. Resistance of pinyon pine to the tip moth (*Dioryctria albovitella*) increases with allozyme heterozygosity. The study was conducted in a natural stand on the lava soils around Sunset Crater, Flagstaff, AZ, which impose both nutrient and water stresses, and allow the levels of herbivores to attain unusually high levels. Resistant trees, R, are indicated by hatched bars, and susceptible trees, S, are indicated by closed bars. Letters above the bars designate statistically significant differences between groups. From Mopper et al (1991).

excess heterozygosity in older or sexually mature trees, I conclude that overdominance is not rare in natural plant populations.

REFERENCES

Bergmann, F., and W. Ruetz. 1991. Isozyme genetic variation and heterozygosity in random tree samples and selected orchard clones from the same Norway spruce populations. For. Ecol. Manage. 46:39–47.

Bergmann, F., and F. Scholz. 1984. Effects of selection pressure by SO_2 pollution on genetic structures of Norway spruce (*Picea abies*). p. 267–275. *In* H.-R. Gregorius (ed.) Population genetics in forestry. Springer-Verlag, Berlin.

Bergmann, F., and F. Scholz. 1987. The impact of air pollution on the genetic structure of Norway spruce. Silvae Genet. 36:80–83.

Bergmann, F., and F. Scholz. 1989. Selection effects of air pollution in Norway spruce (*Picea abies*) populations. p. 143–160. *In* F. Scholz et al. (ed.) Genetic effects of air pollutants in forest tree populations. Springer-Verlag, Berlin.

Boyle, T.J.B., and E.K. Morgenstern. 1986. Estimates of outcrossing rates in six populations of black spruce in central New Brunswick. Silvae Genet. 35:102–106.

Chaisurisri, K., and Y. El-Kassaby. 1994. Genic diversity in a seed production population versus natural populations. Biodiversity Conserv. 3:512–523.

Charlesworth, D. 1991. The apparent selection on neutral marker loci in partially inbreeding populations. Genet. Res. (Cambridge). 57:159–175.

Charlesworth, D., and B. Charlesworth. 1987. Inbreeding depression and its evolutionary consequences. Annu. Rev. Ecol. Syst. 18:237–268.

Cheliak, W.M., and B.P. Dancik. 1982. Genic diversity of natural populations of a clone-forming tree *Populus tremuloides*. Can. J. Genet. Cytol. 24:611–616.

Cheliak, W.M., B.P. Dancik, K. Morgan, F.C.H. Yeh, and C. Strobeck. 1985. Temporal variation and the mating system in a natural population of jack pine. Genetics 109:569–584.

Clarke, B. 1975. The contribution of ecological genetics to evolutionary theory: Detecting the direct effects of natural selection on particular polymorphic loci. Genetics 79:101–108.

Crow, J.F., and M.J. Simmons. 1983. The mutation load in *Drosophila*. p. 1–35. *In* M. Ashburner et al. (ed.) The genetics and biology of *Drosophila*. Vol. 3c. Academic Press, London.

Crumpacker, D.W. 1967. Genetic loads in maize (*Zea mays* L). and other cross-fertilized plants and animals. Evol. Biol. 1:306–423.

Day, D., O.C. De Vos, D. Wilson, and H. Lambers. 1985. Regulation of respiration in the leaves and roots of two *Lolium perenne* populations with two contrasting mature leaf respiration rates and crop yields. Plant Physiol. 78:678–683.

Diehl, W.J. 1988. Genetics of carbohydrate metabolism and growth in *Eisemia foetida* (Oligochatea: Lumbricidae). Heredity 61:379–387.

El-Kassaby, Y., and K. Ritland. 1996. Impact of selection and breeding on the genetic diversity of Douglas-fir. Biodiversity Conserv. 5:795–813.

Endler, J.A. 1986. Natural selection in the wild. Princeton Univ. Press, Princeton, NJ.

Frankel, R. 1983. Heterosis: Reappraisal of theory and practice. Springer-Verlag, Berlin.

Franklin, E.C. 1972. Genetic load in loblolly pine. Am. Nat. 106:262–265.

Frelinger, J.A. 1972. The maintenance of transferrin polymorphisms in pigeons. Proc. Natl. Acad. Sci. USA 69:326–329.

Geburek, Th., F. Scholz, W. Knabe, and A. Vornweg. 1987. Genetic studies by isozyme loci on tolerance and sensitivity in an air polluted *Pinus sylvestris* field trial. Silvae Genet. 36:49–53.

Giannini, R. (ed.). 1991. Effects of pollution on the genetic structure of forest tree populations. *In* Proc. of the meeting Effects of pollution on the genetic structure of forest tree populations., Rome. 3 Apr. 1990. Natl. Res. Council of Italy, Firenze.

Ginzburg, L.R. 1979. Why are heterozygotes often superior in fitness? Theor. Popul. Biol. 15:264–267.

Ginzburg, L.R. 1983. Theory of natural selection and population growth. Benjamin/Cummings Publ. Co., Menlo Park, CA.

Goncharenko, G.G., V.E. Padutov, and A.E. Silin. 1993a. Allozyme variation in natural populations of Eurasian pines: I. Population structure, genetic variation, and differentiation in *Pinus pumila* (Pall.) regel from Chukotsk and Sakhalin. Silvae Genet. 42:237–246.

Goncharenko, G.G., V.E. Padutov, and A.E. Silin. 1993b. Allozyme variation in natural populations of Eurasian pines: II. Genetic variation, diversity, differentiation and gene flow in *Pinus sibirica* Du Tour in some lowland and mountain populations. Silvae Genet. 42:246–253.

Goncharenko, G.G., A.E. Silin, and V.E. Padutov 1995. Allozyme variation in natural populations of Eurasian pines: III. Populations structure, diversity, differentiation and gene flow in central and isolated populations of *Pinus sylvestris* L. in Eastern Europe and Siberia. Silvae Genet. 43:119–132.

Hall, J.G., and R.K. Koehn. 1983. The evolution of enzyme catalytic efficiency and adaptive inference from steady-state kinetic data. Evol. Biol. 16:53–96.

Hamrick, J.L., and M.J. Godt. 1990. Allozyme diversity in plant species. p. 43–63. *In* A.H.D. Brown et al. (ed.) Plant population genetics, breeding and genetic resources. Sinauer Associates, Sunderland, MA.

Hedrick, P.W., and O. Savolainen. 1996. Molecular and adaptive variation: A perspective for endangered plants. p. 92–102. *In* J. Maschinski et al. (ed.) Southwestern rare and endangered plants. Proc. of the 2nd Conf., Flagstaff, AZ. 11–14 Sept. 1995. Gen. Tech. Rep. RM-GTR-283. USDA-FS, Rocky Mountain For. and Range Exp. Stn., Fort Collins, CO.

Jelinski, D.E., and W.M. Cheliak. 1992. Genetic diversity and spatial subdivision of *Populus tremuloides* (Salicaceae) in a heterogeneous landscape. Am. J. Bot. 79:728–736.

Kertadikara, A.W.S., and D. Prat. 1995. Genetic structure and mating system in teak (*Tectona grandis* L. f.) provenances. Silvae Genet. 44:104–110.

Kimura, M., and T. Ohta. 1971. Theoretical aspects of population genetics. Princeton Univ. Press, Princeton, NJ.

Koehn, R.K. 1978. Physiology and biochemistry of enzyme variation: The interface of ecology and population genetics. p. 51–71. *In* P. Brussard (ed.) Ecological genetics: The interface. Springer-Verlag, New York.

Koehn, R.K., and S.E. Shumway. 1982. A genetic/physiological explanation for differential growth rate among individuals of the American oyster, *Crassostrea virginica* (Gmelin). Mar. Biol. Letters 3:35–42.

Koehn, R.K., A.J. Zera, and J.G. Hall. 1983. Enzyme polymorphism and natural selection. p. 115–136. *In* M. Nei and R.K. Koehn (ed.) Evolution of genes and proteins. Sinauer Associates, Sunderland, MA.

Lande, R., and D.W. Schemske. 1985. The evolution of self fertilization and inbreeding depression in plants: I. Genetic models. Evolution 39:24–40.

Lerner, I.M. 1954. Genetic homeostasis. Oliver & Boyd, Edinburgh.

Lesica, P., and F.W. Allendorf. 1992. Are small populations of plants worth preserving? Conserv. Biol. 6:135–139.

Lewandowski, A., J. Burczyk, and L. Meinartowicz. 1991. Genetic structure and the mating system in an old stand of Polish larch. Silvae Genet. 40:75–79.

Linhart, Y.B., J.B. Mitton, K.B. Sturgeon, and M.L. Davis. 1981. Genetic variation in space and time in a population of ponderosa pine. Heredity 46:407–426.

Meinartowicz, L., and A. Lewandowski. 1994. Allozyme polymorphisms in seeds collected from a IUFRO-68 Douglas-fir test-plantation. Silvae Genet. 43:181–186.

Mitchell-Olds, T. 1995. Interval mapping of viability loci causing heterosis in *Arabidopsis*. Genetics 140:1105–1109.

Mitton, J.B. 1989. Physiological and demographic variation associated with allozyme variation. p. 127–145. *In* D. Soltis and P. Soltis (ed.) Isozmes in plant biology. Dioscorides Press, Portland, OR.

Mitton, J.B. 1997. Selection in natural populations. Oxford Univ. Press, New York.

Mitton, J.B., and M.C. Grant. 1984. Associations among protein heterozygosity, growth rate, and developmental homeostasis. Annu. Rev. Ecol. Syst. 15:479–499.

Mitton, J.B., and M.C. Grant. 1995. Genetics and the natural history of quaking aspen. BioScience 46:25–31.

Mitton, J.B., and R.M. Jeffers. 1989. The genetic consequences of mass selection for growth rate in Engelmann spruce. Silvae Genet. 38:6–12.

Mopper, S., J.B. Mitton, T.G. Whitham, N.S. Cobb, and K.M. Christensen. 1991. Genetic differentiation and heterozygosity in pinyon pine associated with resistance to herbivory and environmental stress. Evolution 45:989–999.

Mopper, S., and T.G. Whitham. 1986. Natural bonsai of Sunset Crater. Nat. Hist. 95:42–47.

Morgante, M., G.G. Vendramin, P. Rossi, and A.M. Olivieri. 1993. Selection against inbreds in early life-cycle phases in *Pinus leucodermis* Ant. Heredity 70:622–627.

Muller-Stark, G. 1985. Genetic differences between "tolerant" and "sensitive" beeches (*Fagus sylvatica* L.) in an environmentally stressed adult forest stand. Silvae Genet. 34:241–247.

Neale, D.B., and W.T. Adams. 1985. Allozyme and mating-system variation in balsam fir (*Abies balsamea*) across a continuous elevational transect. Can. J. Bot. 63:2448–2453.

Noltman, E.A. 1972. Aldos-ketose isomerases. p. 271–354. *In* P.D. Boyer (ed.) Enzymes. Vol. 6. Academic Press, New York.

Parsons, P.A. 1971. Extreme environment heterosis and genetic loads. Heredity 26:479–483.

Parsons, P.A. 1973. Genetics of resistance to environmental stresses in *Drosophila* populations. Annu. Rev. Genet. 7:239–265.

Parsons, P.A. 1987. Evolutionary rates under environmental stress. Evol. Biol. 21:311–347.

Plessas, M.E., and S.H. Strauss. 1987. Allozyme differentiation among populations, stands, and cohorts in Monterey pine. Can. J. For. Res. 16:1155–1164.

Politov, D.V., and K.V. Krutovskii. 1994. Allozyme polymorphism, heterozygosity, and mating system of stone pines. p. 36–42. *In* W.C. Schmidt and F.-K. Holtmeier (ed.) Proc. Int. Workshop on Subalpine Stone Pines and their Environment: The Status of Our Knowledge. USDA For. Serv. Gen. Tech. Rep. INT-GTR-309. USDA-FS, Intermountain Res. Stn., Ogden, UT.

Powers, D.A., M. Smith, I. Gonzalez-Villasenor, L. DiMichele, D. Crawford, G. Bernardi, and T. Lauerman. 1994. A multidisciplinary approach to the selection/neutralist controversy using the model teleost *Fundulus heteroclitus*. p. 43–107. *In* D. Futuyma and J. Antonovics (ed.) Oxford surveys in evolutionary biology. Vol. 9. Oxford Univ. Press, Oxford.

Prus-Gowacki, W., and St. Godzik. 1995. Genetic structure of *Picea abies* trees tolerant and sensitive to industrial pollution. Silvae Genet. 44:62–65.

Rainey, D.Y., J.B. Mitton, and R.K. Monson. 1987. Associations between enzyme genotypes and dark respiration in perennial ryegrass, *Lolium perenne* L. Oecologia 74:335–338.

Rainey, D.Y., J.B. Mitton, R.K. Monson, and D. Wilson. 1990. Effects of selection for dark respiration rate on enzyme genotypes in *Lolium perenne* L. Ann. Bot. (London) 66:649–654.

Rainey-Foreman, D., and J.B. Mitton. 1995. Glucose utilization by 6pgd genotypes in *Lolium perenne*. Molec. Ecol. 4:231–237.

Rothe, G.M., and F. Bergmann. 1995. Increased efficiency of Norway spruce heterozygous phosphoenolpyruvate carboxylase phenotype in response to heavy air pollution. Angew. Bot. 69:27–30.

Scholz, F., and F. Bergmann. 1984. Selection pressure by air pollution as studied by isozyme-gene-systems in Norway spruce exposed to sulphur dioxide. Silvae Genet. 33:238–241.

Scholz, F., H.-R. Gregorius, and D. Rudin (ed.). 1989. Genetic effects of air pollutants in forest tree populations. Springer-Verlag, Berlin.

Scott, T.M., and R.K. Koehn. 1990. The effect of environmental stress on the relationship of heterozygosity to growth rate in the coot clam *Mulinia lateralis* (Say). J. Exp. Mar. Biol. Ecol. 135:109–116.

Shaw, D.V., and R.W. Allard. 1982. Isozyme heterozygosity in adult and open-pollinated embry samples of Douglas-fir. Silva. Fenn. 16:115–121.

Snyder, L.R.G. 1981. Deer mouse hemoglobins: Is there genetic adaptation to high altitude? BioScience 31:299–304.

Sorensen, F., and R.S. Miles. 1982. Inbreeding depression in height, height growth, and survival of Douglas-fir, ponderosa pine, and noble fir to 10 years of age. For. Sci. 28:283–292.

Shull, G.H. 1948. What is "heterosis"? Genetics 33:439–446.

Simmons, M.J., and J.F. Crow. 1977. Mutations affecting fitness in *Drosophila* populations. Annu. Rev. Genet. 11:49–78.

Smouse, P.E. 1986. The fitness consequences of multiple-locus heterozygosity under the multiplicative overdominance and inbreeding depression models. Evolution 40:946–957.

Templeton, A.R. 1982. Adaptation and the integration of evolutionary forces. p. 15–31. *In* R. Milkman (ed.) Perspectives on evolution. Sinauer Associates, Sunderland, MA.

Teska, W.R., M.H. Smith, and J.M. Novak. 1990. Food quality, heterozygosity, and fitness correlates in *Peromyscus polionotus*. Evolution 44:1318–1325.

Turelli, M., and L. Ginzburg. 1983. Should individual fitness increase with heterozygosity? Genetics 104:191–209.

Wallace, B. 1959. Studies of the relative fitnesses of experimental populations of *Drosophila melanogaster*. Am Nat. 93:295–314.

Wallace, B. 1981. Basic population genetics. Columbia Univ. Press, New York.

Whitham, T.G., and S. Mopper. 1985. Chronic herbivory: Impacts on architecture and sex expression of pinyon pine. Science (Washington, DC) 228:1089–1091.

Wilson, D. 1982. Response to selection for dark respiration rate of mature leaves in *Lolium perenne* and its effects on growth of young plants and similar simulated swards. Ann. Bot. (London) 49:303–312.

Wilson, D., and J.G. Jones. 1982. Effect of selection for dark respiration rate of mature leaves on crop yields of *Lolium perenne* cv s23. Ann. Bot. (London) 49:313–320.

Wright, S. 1977. Evolution and genetics of populations. Vol. 3. Experimental results and evolutionary deductions. Univ. of Chicago Press, Chicago.

Yeh, F.C., M.A.K. Khalil, Y.A. El-Kassaby, and D.C. Trust. 1986. Allozyme variation in *Picea mariana* from Newfoundland: Genetic diversity, populations structure, and analysis of differentiation. Can. J. For. Res. 16:713–720.

Zera, A.J., R.K. Koehn, and J.G. Hall. 1983. Allozymes and biochemical adaptation. p. 633–674. *In* G.A. Kerkut and L.I. Gilbert (ed.) Comprehensive insect physiology, biochemistry and pharmacology. Pergamon Press, New York.

6 Role of Chromosome Blocks in Heterosis and Estimates of Dominance and Overdominance[1]

Edwin T. Bingham

University of Wisconsin
Madison, Wisconsin

ABSTRACT

Chromosome blocks are the genomic units of genetic transmission in sexual reproduction. We work with chromosome blocks, not individual genes in our conventional breeding and genetic research. Thus, chromosome blocks underpin heterosis and estimates of gene action. Chromosome blocks vary in size according to intensity of linkage (frequency of recombination) and the number of sexual generations (the approach to linkage equilibrium); however, all we usually know is the number of sexual generations. Even in the transfer of *single gene traits* by backcrossing, we usually do not know how much genetic material is linked to the gene of interest. The latter is sometimes referred to as *linkage drag*. D.F. Jones clearly recognized the role of chromosome blocks in 1917 when he proposed dominance of linked factors as a means of accounting for heterosis. The proposition is elegant because it acknowledges the cumulative effect of linked dominant genes as transmission units. In the years to follow there was much debate about gene action, and heterosis was sometimes interpreted as true overdominance—single loci at which the heterozygous phenotype exceeds that of either homozygote. Maize (*Zea mays* L.) researchers were careful to point out that estimates of dominance variance exceeding that for straight dominance could be due to either overdominance or linkage disequilibrium of linked loci with favorable alleles in repulsion phase (pseudo-overdominance). Maize researchers went on to compare degrees of dominance in F_2 populations in linkage disequilibrium with populations in F_8 through F_{16} in linkage equilibrium. Estimates for degree of dominance were reduced with the approach to linkage equilibrium indicating that the initial heterosis was more likely due to Jones' dominance of linked factors in linkage disequilibrium, than due to true overdominance. In autotetraploid alfalfa, we reached the same conclusion from results indicating dominant linked factors in chromosome blocks, and not multiple allelic interactions, explained improvement and maximum heterosis. Currently, molecular-marker-facilitated investigations of quantitative trait loci in maize report often finding higher yield in the heterozygote than in either homozygote. Based on past research and the fact that chromosome blocks are the units of sexual transmission, it seems likely that the bulk of these heterozygote effects are due to dominance of linked factors as proposed by Jones. Dominant alleles at different loci complement each other by masking recessive alleles at

---

[1] Contribution from the Dep. of Agronomy, Univ. of Wisconsin Agric. Exp. Stn., Madison. Research supported by the College of Agricultural and Life Sciences, University of Wisconsin, Madison.

Copyright © 1998 Crop Science Society of America, 677 S. Segoe Rd., Madison, WI 53711, USA. *Concepts and Breeding of Heterosis in Crop Plants.* CSSA Special Publication no. 25.

respective loci. The gene action is nonallelic gene interaction or epistasis. Finally, the cumulative action of genes in chromosome blocks not only explains the breeding behavior of cross-pollinated crops, but also explains the fixation of transgressive traits in self-pollinated crops, and the ability of auto- and allopolyploids to conceal deleterious recessive traits.

Plants have three genomes: nuclear, plastid, and mitochondrial; however, most of the DNA among the three genomes is organized on chromosomes in the nucleus. This discussion of heterosis will concern genes on chromosomes and the trait of interest will be yield. Probably all will agree that yield usually is a complex trait controlled by many genes. The effects of some genes will be large enough to identify as quantitative trait loci (QTL) that cosegregate with molecular markers. The effects of other genes will be small, but have a cumulative effect on yield. In any case, once it is agreed that many genes are involved, genetic linkages are bound to exist. Linkages define the chromosome blocks in this discussion. Our knowledge of genetics and sexual reproduction indicates that the chromosome blocks are the units of genetic transmission that underpin heterosis, and estimates of dominance and overdominance.

The book *Heterosis* edited by Gowen (1952) contains chapters by the pioneers in heterosis, and was very useful in preparing this review. The book is based on presentations at the benchmark conference on Heterosis held at Iowa State in the summer of 1950. More recently, the genetic basis of heterosis has been reviewed by Brewbaker (1964), Jinks (1983), Sprague (1983), and Stuber (1994). A review of the biochemical and physiological basis of heterosis can be found in an excellent article by Rhoades et al. (1992).

REALITY OF CHROMOSOME BLOCKS

The number of chromosome blocks is a function of the chromosome number and the number of crossovers per chromosome arm during meiosis. Both factors vary among crop species, but in all species there is a relatively small number of chromosome blocks or units of genetic transmission relative to the amount of nuclear DNA and presumed number of genes. In general, there is only one to two crossovers per chromosome arm per meiocyte. The positions of crossing over will differ in each meiocyte; hence the linkages also will differ, but the number of chromosome blocks will be relatively similar among the gametes produced by each meiocyte. In homozygous inbred lines the position of crossing over has no effect on linkages or the genotype of the gamete. On the other hand, in heterozygous materials both the linkage order and genotype of the gamete are affected by the position of crossing over. Thus, F_1 hybrid plants with the same genotype will produce an array of different gametes in linkage disequilibrium based on different positions of crossing over in different meiocytes.

How many chromosome blocks do we work in heterozygous or segregating parents in our breeding programs? In maize, if we assume about three crossovers per chromosome pair, there would be 3 blocks × 10 haploid chromosomes, or about 30 units of genetic transmission per gamete. In alfalfa (*Medicago sativa* L.)

if we assume about 2 crossovers per chromosome pair there would be about 2 blocks × 16 haploid chromosomes, or about 32 transmission units. The exact number is not important because the number of units of genetic transmission will be small compared with the number of genes in the genome in any species.

DEFINITIONS

Additive, dominant and overdominant gene action are essentially defined in Fig. 6–1. An important consideration in discussing gene action is that an allele with complete dominance also has an additive effect. Moreover, these additive effects are cumulative, thus giving rise to the often used expression: heterosis is due to the cumulative effects of favorable alleles with partial to complete dominance (Hallauer & Miranda, 1988).

Genetic equilibrium (the Hardy-Weinberg Law) is when the relative frequencies of each allele tend to remain constant from generation to generation in the absence of mutation, selection, random drift, and migration. In diploids, genetic equilibrium is reached after one generation of random mating, e.g., in the F_2 or Syn-1 generation. Linkage equilibrium on the other hand, approaches a practical state about F_8 and approaches the real state at about F_{16}, as we will see in the case of maize discussed later. The difference between genetic equilibrium and linkage equilibrium is central to this discussion of chromosome blocks.

In autotetraploids such as alfalfa and potato (*Solanum tuberosum* L.), genetic equilibrium is approached asymptotically in about the F_{12} generation. Linkage equilibrium then becomes a theoretical state that would require more generations than possible in most breeding programs. Historically, chromosome blocks have been termed effective factors by Mather and Jinks (1973), and linkats by Demarly (1979). The term effective factors is generally used in the literature where researchers have estimated the number of effective factors controlling certain traits (Dudley et al., 1974). In alfalfa we have used linkats when discussing chromosome blocks (Bingham et al., 1994).

Repulsion phase linkage, linkage disequilibrium, linkage bias, and pseudo-overdominance are used interchangeably in the literature. The terms all describe the model of a chromosome block with dominance and recessive linked loci in

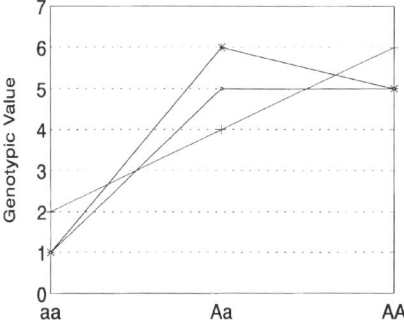

Fig. 6–1. Genotypic values at a theoretical locus with alleles A and a with additive (+), dominant (□). and overdominant (*) gene action.

A	a	A	a
b	B	B	b
C	c	c	C
d	D	d	D
Repulsion Phase Linkage		Coupling Phase Linkage	

Fig. 6–2. Theoretical models of chromosome blocks in repulsion and coupling phase linkage, respectively.

Fig. 6–2. Similarly, coupling phase linkage (Fig. 6–2) also can contribute to linkage disequilibrium, linkage bias and pseudo-overdominance. Coupling phase linkage is often discussed in self-pollinated crops and in the case of improved inbreds of cross-pollinated crops. In either coupling or repulsion phase linkage, the dominant alleles at different loci complement each other by masking recessive alleles at respective loci. This complementary gene interaction is nonallelic gene interaction or epistasis.

HISTORICAL PERSPECTIVE

Heterosis was first used by Shull, as a synonym for hybrid vigor, and was not intended to suggest any mechanism (Shull, 1952). Heterosis is perhaps the greatest genetic phenomenon in nature that can be exploited without understanding it. Thus, a great seed industry is based on it, while we are continuing to do research to understand the genetic mechanisms of heterosis.

There have been two explanations of heterosis beginning with East (1908), Shull (1908, 1911), and East and Hayes (1912), all of whom believed that different germplasms produce a developmental stimulus that increases with the diversity (heterozygosity) of the uniting gametes. This is now called the overdominance hypothesis. Under this interpretation, the heterozygote has an advantage (Hull, 1952). Alternatively, heterosis can be produced by the masking of deleterious recessive alleles in each strain by dominant, or nearly dominant alleles from the other strain. This is the dominance hypothesis. Davenport (1908) was the first to emphasize that the effects of deleterious recessive genes tend to be concealed in heterozygotes by dominance. Bruce (1910) and Keeble and Pellew (1910) appear to be the first to explicitly state the dominance hypothesis.

Objections to the dominance hypothesis between 1910 and 1917 were that selection in maize failed to produce inbred lines as good as hybrids (Shull, 1911; East & Hayes, 1912), and that there was an absence of a skewed F_2 distribution expected from the expansion of 3/4 dominants + 1/4 recessives (Emerson & East, 1913). These objections were largely dispelled when Jones (1917) pointed out that with close linkage, and Collins (1921) found that with a large number of factors even without linkage, the two hypotheses became indistinguishable. Thus,

```
           INBRED PARENT 1                        INBRED PARENT 2
            CHROMOSOME                              CHROMOSOME
    I         II         III              I          II         III

   A A       m m        W W              a a        M M        w w
   b b       N N        x x              B B        n n        X X
   C C       o o        Y Y              c c        O O        y y
   d d       P P        z z              D D        p p        Z Z

        6 loci with dominant alleles         6 loci with dominant alleles

                           F₁ HYBRID

                    A a         m M        W w
                    b B         N n        x X
                    C c         o O        Y y
                    d D         P p        z Z

                        12 loci with dominant alleles
```

Fig. 6–3. Models of linked genes on chromosome blocks patterned after the model by Jones (1917). An inbred with six loci with dominant alleles linked in repulsion, when crossed with an inbred with six different loci with dominant alleles, produces an F_1 hybrid with 12 loci with dominant alleles.

what appeared to be overdominance could be explained by linked dominant factors.

The proposition of D.F. Jones (1917) that dominance of linked factors could account for heterosis is a benchmark. Jones' model is presented in Fig. 6–3, using inbred parents and three hypothetical chromosomes. Each chromosome has two dominant and two recessive loci linked in repulsion, it is assumed that one inbred has one set of dominant alleles and the other inbred has a different set in a complementary repulsion-phase linkage. In total, one inbred has six loci with dominant alleles for yield, as an example, and the other inbred has a different set of six. In the F_1, there is a dominant at each locus for a complementary set of 12 loci with a dominant allele, and a hypothetical F_1 yield double that of either inbred. Inbreds of maize that were only one-half as productive as the hybrid were common at the time.

Jones went on to discuss how with relatively tight linkage of dominant and recessive alleles it would be difficult to recombine in any one individual in later generations any greater number of dominants in the homozygous condition than were present in the parents. This disarmed the criticism about not being able to produce pure breeding maize lines as high in yield as the F_1. Following the hypothetical case into the F_2 generation by selfing or sib-mating, he demonstrated that linkage of the dominant factors in repulsion produced a normal distribution of hypothetical yield based on the number of loci with a dominant allele. This disarmed the second main criticism of the dominant hypothesis in that no skewedness was expected with linkage. The final point in Jones' paper anticipated the eventual population improvement in maize. He pointed out that recombination in this dominant linked factor model provided a means of understanding how certain homozygous individuals (and varieties) may possess a greater number of desirable characters than others.

Crow (1993) noted that Sewall Wright published a series of three papers in the 1920s on the effects of inbreeding and crossing in guinea pigs that are important in the history of heterosis. Wright noted the decline in vigor with inbreeding; the fixation of different traits in different lines; the immediate recovery of vigor when inbred lines are crossed; and the predictable decline when hybrids are inbred. He stated that although these results are all direct consequences of Mendelian inheritance and dominance, they are equally consistent with partial dominance or overdominance.

In the late 1940s there was a resurgence of interest in idea that heterozygosity as such was important in heterosis. In plants, this was largely because of maize studies by Hull (see Hull, 1952). He introduced the word overdominance to describe the situation. Hull (1952) notes that Fisher (1918, 1931) used the term *super-dominance*, but that the term had never caught on. At the conference on heterosis, Hull (1952) listed eight factors as evidence of overdominance in the genetics of grain yield of maize. The list appears very dated now because of successful population improvement in maize, and because of significant inbred and hybrid improvement since 1950. Nevertheless, Hull's list is historically important and presented here in its entirety.

Evidence of overdominance in the genetics of grain yield of maize consists of:

1. Failure of mass selection and ear-to-row selection beyond the level of the adapted variety.
2. Crossbreeding recombinations of parent lines of elite hybrids yield little more than the original varieties.
3. Hybrids of second-cycle and third-cycle lines yield little more than those of the first cycle.
4. Homozygous maize yields 30% as much as heterozygous maize.
5. No evidence of epistasis in maize yield.
6. Regression analyses of yields of F_1s and inbred parents indicate a zone of nearly level regression near the upper end of the range of present data, where it might be predicted with the kind of artificial selection which has been practiced, and in the event of overdominance.
7. There is some evidence that selection for general combining ability alone with respect to yield is effective and this too is consistent with the expectation of overdominance theory.
8. The fact of hybrid maize is hardly to be explained as other than a result of selection for specific combinability, which in turn is manifestly dependent on heterozygosity of maize yield genes.

In the same period of the late 1940s through the 1950s, overdominance was championed by some fruit fly geneticists. Crow (1993) presents an engrossing chronology of the geneticist personalities and their views. Crow's article provides a thorough treatment of the evolution of thought patterns about heterosis by Crow and others, and he relates heterosis to mutation, mean fitness, and genetic load. At the time of the heterosis conference in 1950, Crow (1952), Dobzhansky (1952), and others believed that although overdominant loci might be exceptional, they were of disproportionate influence in maintaining a stable equilibrium with intermediate allele frequencies. By 1955, according to Crow, Dobzhansky changed from the view that overdominant loci were exceptional, though important, to the view that they were ubiquitous.

Muller disagreed with Dobzhansky, and the details are best obtained from Crow's article (1993). Evidently, Muller's emphasis on partial dominance and appropriate additivity between loci led him to believe that selection is an efficient process. He noted the success of breeding programs in plants and animals, and believed that a positive eugenics program would be effective. Dobzhansky, believing in overdominance, contended that selection would be complicated and unpredictable. Crow (1993) points out that indeed selection under an overdominance model would be complicated, because, Haldane had shown that at equilibrium with overdominance, the parent offspring correlation in fitness is zero.

Crow (1993) noted that the arguments that he presented in 1948 for overdominance in maize yield had not held up with newer data. He cites the articles from Nebraska and North Carolina on the decrease in dominance over generations with the approach to equilibrium. These articles are reviewed in the following section of this chapter. Crow further notes that relatively high yielding maize inbreds have now been developed. They are not yet as good as the best hybrids but they are better than would be expected if there were a large contribution from overdominance. Thus, in maize, Crow agrees with the interpretation of pseudo-overdominance due to linkage disequilibrium.

DECLINE IN EVIDENCE FOR OVERDOMINANCE IN MAIZE

There was a scholarly ferment about heterosis around 1950 judging from the literature and the articles delivered at the heterosis conference (Gowen, 1952). It was a most timely conference because the pioneers in the concept and terminology were able to record their thoughts and conclusions in person. From this ferment came several important articles on mating systems and procedures to estimate the degree of dominance controlling quantitative characters (Comstock & Robinson, 1952; Robinson et al., 1949; Gardner et al., 1953). Estimates of the degree of dominance in open-pollinated populations were in the range of partial dominance (Robinson et al., 1955). On the other hand, estimates based on the F_2 generation of a single cross of two inbreds were in the range of overdominance (Robinson et al., 1949; Gardner et al., 1953).

The authors were careful to point out that it was theoretically possible to obtain estimates of overdominance due to repulsion phase linkages in their material even though the individual genes may have no more than partial dominance. Moreover, they noted that advanced generations of hybrid populations obtained by random mating and approaching linkage equilibrium could be used to determine whether linkage was in fact an important source of bias in the estimates. This set the stage for definitive experiments to follow at North Carolina State University and the University of Nebraska.

The strategy was to advance the generations by random mating toward linkage equilibrium with an opportunity for free recombination between successive generations. This should allow linkages between loci to break up. Linked chromosome blocks containing loci with dominant and recessive alleles would mimic overdominance because of the cumulative effect of the dominants, and their break up would reduce the overdominant effect on the average.

In Nebraska, results comparing F_2 and F_8 populations of maize in linkage disequilibrium and approaching equilibrium respectively, involved a cross of inbreds M14 and 187-2 (Gardner & Lonnquist, 1959). The inbreds were considered typical of lines used in the Corn Belt at the time. The strategy, used the North Carolina Design III, which produced the F_2 by selfing the F_1, and the F_8 by allowing random pollination in large isolated blocks each generation. Then, randomly chosen F_2 and F_8 plants were used as male parents and each was crossed to each inbred parent to produce pairs of backcross progenies from respective male parents. One hundred such pairs were tested at two locations. Results from the two locations differed somewhat but still allowed the following general conclusions. The degree of dominance and the dominance variance were lower for every quantitative trait studied in the F_8 generation compared with the F_2. This indicated that estimates of overdominance in the F_2 were biased upward as a result of repulsion phase linkages many of which were broken by the F_8 generation. In the F_8 generation, estimates of dominance were no more than partially or completely dominant. Interestingly, there were consistently high additive genetic variances in the F_8 generation compared with the F_2 generation.

In 1960, estimates of dominance in F_2 versus F_8 generations derived from two North Carolina hybrids were published (Robinson et al., 1960). Once again estimates of dominance decreased with the approach to linkage equilibrium in the F_8 in both populations. As in the Nebraska study, this indicated that estimates of overdominance in the F_2 were biased upward as a result of linkage bias. The authors noted that dominance variance if biased by linkage is expected to be reduced following recombination regardless of the type of linkage. Moll and Robinson (1967) published four more estimates of dominance on North Carolina materials taken to F_8, F_{12}, and F_{13}. Average level of dominance was decreased in the advanced generation in all cases. In the North Carolina materials, additive variance also decreased over generations of recombination, although the decreases were less than those for dominance. This permitted some interesting interpretations. The authors indicated that there might be no change in additive genetic variance following recombination if the repulsion and coupling linkages tended to balance each other. An increase in additive variance could occur with predominantly repulsion phase linkages in the parent lines. Recall that this is what happened in the Nebraska material (Garnder & Lonnquist, 1959).

With initial coupling-phase linkages, the additive genetic variance would be expected to decrease, the rate of decrease being dependent upon the predominance of the coupling linkages. Hence, the North Carolina materials must have contained a preponderance of coupling phase linkages. Coupling phase linkages are thought to be indicative of improvement and are commonly referred to in self-pollinated crops with pure line varieties.

The capstone on the estimates of dominance in the Nebraska population is at generation F_{16} (Fig. 6–4). The values represent an average of those from Lonnquist (1980), and from Gardner (1992, personal communication) in a similar figure (Crow, 1993). Tests at different locations gave slightly different values,

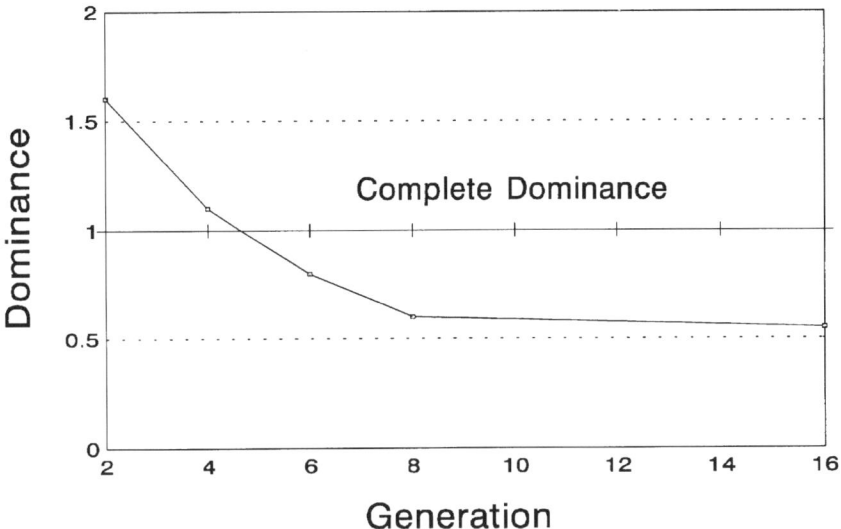

Fig. 6–4. Summary of results showing a decrease in estimates of dominance in maize with the approach to linkage equilibrium (Gardner & Lonnquist, 1959; Lonnquist, 1980; Gardner, 1992, personal communication).

but the trends in the estimates of dominance and the conclusions are the same. The estimates of dominance decreased with the approach to linkage equilibrium. The decrease between F_8 and F_{16} was small, indicating that F_8 estimates are practical and realistic. Both F_8 and F_{16} estimates of dominance are in the range of partial dominance.

SHIFT FROM OVERDOMINANCE TO CHROMOSOME BLOCKS IN ALFALFA

Alfalfa behaves as an autotetraploid and heterosis progresses to a maximum one or two generations after a single cross (Bingham et al., 1994). Progressive heterosis is due to complex tetrasomic segregations. Researchers studying progressive heterosis always noted that they could not distinguish between linked chromosome blocks and multiple alleles at a locus producing overdominance in the progressive heterosis phenomenon (Demarly 1963; Dunbier & Bingham, 1975; Groose et al., 1989); however, between 1965 and 1975 heterosis in alfalfa was often discussed in terms of overdominance. About 1975 researchers began long-term research to separate the effects of chromosome blocks from multiple alleles. In this research, cultivated autotetraploid alfalfa was haploidized to produce cultivated diploids, and diploid hybrids were chromosomally doubled to produce two-allele autotetraploids that could not contain more than one interaction of two alleles. Multiple allelic interactions of three or four alleles at a locus were eliminated. Two allele populations were produced from single plants by first selfing, and then sib-mating. Sib-mating and selection were

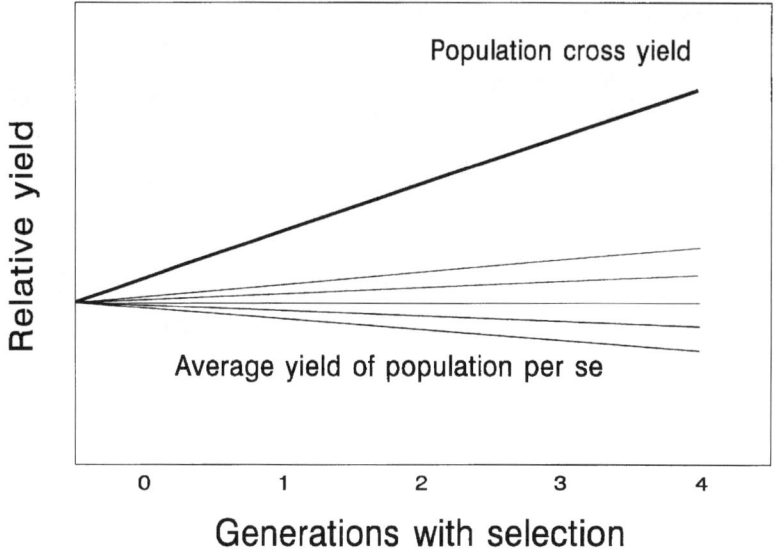

Fig. 6–5. Yield improvement in population crosses of alfalfa following generations of inbreeding and selection in individual populations.

practiced in subsequent generations of the two allele populations. Population improvement for yield was realized in a set of experiments (Pfeiffer & Bingham, 1983) and another set a decade later (Woodfield & Bingham, 1995; Fig. 6–5). Improvement in crosses was explained by accumulation of favorable alleles with additive to completely dominant effects. Improvement took place over generations of recombination between the chromosome blocks brought together in the original diploid hybrid. Allelic interactions or overdominance could not increase in this breeding strategy, thus implicating the theory that the accumulation of favorable alleles on chromosome blocks and not overdominance in alfalfa was responsible for the yield gains.

HETEROSIS IN SINGLE CHROMOSOME BLOCKS

At the 1950 Heterosis Conference, Jones (1952) described his backcrossing experiments involving single chromosome blocks. Several deep inbred lines of maize, one of which was in the 41st generation of continuous self-fertilization, were crossed to unrelated inbred lines carrying dominant marker genes. The markers were red cob, yellow endosperm, and non-glossy seedlings. They were chosen because they had little or no effect on plant growth. The strategy was to backcross the markers into the inbreds, which at BC6 essentially restores the original inbred condition, except for the chromosome block containing the dominant marker and its linked genes. Segregates in BC6 with the dominant marker are heterozygous for the chromosome block, while normal segregates are homozygous for the inbred condition. Thus, any heterotic effects of that single chromosome block can be studied. Jones measured the effect of each block on height. Differences were small, but nearly all indicated a heterotic effect on height. He considered the three marked chromosome blocks as samples of typical chromosome blocks, and concluded that there are genes in all parts of the maize chromosomes that contribute to normal growth and development.

The same concept of studying the effects of chromosome blocks from teosinte in an inbred of maize was discussed by Mangelsdorf (1952). He carried out four generations of backcrossing teosinte into an inbred of maize, and measured the effects of the added chromosome blocks on the inbreds traits and performance. Introgressed segments involved blocks of genes because the segments had multiple effects. In recent research on the origin of maize Doebley et al. (1995) have transferred teosinte blocks into maize and maize blocks into teosinte using reciprocal backcrossing. The strategy of studying single chromosome blocks has provided insight on the evolution of epistasis and dominance in maize.

Lamkey et al. (1988) discussed the importance of chromosome blocks in a study of the contribution of the long arm of maize chromosome 10 to heterosis. They used the ingenious method of B-A translocations to transfer intact chromosome segments from one inbred line of maize to another, and then study the heterotic effect of the segment in an otherwise inbred background (Robertson, 1964; Peterson & Wernsman, 1964). In the study by Lamkey et al. (1988) the long arm of chromosome 10 is only about 3% of the genome, but accounted for 9.4 to 20.8% of total heterosis.

Burton (1986) identified chromosome blocks in the heterozygous condition in isogenic populations of pearl millet [*Pennisetum glaucum* (L.) R.Br.] in several creative ways. He demonstrated the existence of heterosis due to single heterozygous chromosome blocks and suggested how they could be used to increase the forage yield of a top yielding hybrid. For example, Burton and Werner (1991) backcrossed conventional genetic markers into inbreds of pearl millet and found significant heterotic effects of single blocks. Burton and Wilson (1995) described a modification of the method to screen for heterozygous chromosome blocks in tropical land races of pearl millet, that can replace the block in the male parent of a top producing F_1 hybrid and increase forage yield. The method works with chromosome blocks in which a single visibly selectable marker is located. Thus, only a small amount of the genome is evaluated for new chromosome blocks in that region, yet measurable heterosis is found in the single chromosome blocks of some land races. Currently, molecular markers can be used in marker assisted selection, and the transfer of single blocks for plant improvement is expected to increase.

HETEROSIS ASSOCIATED WITH QUANTITATIVE TRAIT LOCI

The benchmark hybrid of maize inbreds B73 × Mo17 has been used to identify QTL contributing to heterosis (Stuber, 1992; Stuber et al., 1992). A modified Design III and molecular markers were used to identify QTL in the F_2–F_3 when the hybrid population was still in linkage disequilibrium. Whenever a QTL for grain yield was detected, the heterozygote had a higher phenotype than the respective homozygote, with only one exception. It seems likely, based on the previous results in maize where the level of dominance decreased with the approach to linkage equilibrium, that pseudo-overdominance due to linkage disequilibrium is involved in the heterozygotes in early generations.

Cockerham and Zeng (1996) extended Design III to include linkage, two locus epistasis, and the use of F_3 data, and applied it to Stuber's maize data. Their analysis strongly suggested that there are multiple linked QTL in many chromosomes. They pointed out that when several QTL are linked, with an aggregation of dominance effects, pseudo-overdominance can be created and observed in the single marker analysis. The interesting test will be to random mate the population to at least F_8, and compare the levels of dominance in the two populations. Even after separation of as many linked loci as possible by recombination there still may be clustering of some genes contributing to a QTL effect.

Khavkin and Coe (1995a,b) found evidence of clustering of genes for growth and development in maize. They surveyed data from naked-eye polymorphisms and published molecular marker data, and found functional clusters of genes distributed nonrandomly along all 10 chromosomes. They presume that clusters are functional units of genes expressed in concert and affecting plant development. Major QTL for plant height, earliness, and grain yield are visible manifestations of developmental clusters. It will not be surprising if such clusters turn out to be tightly or completely linked on chromosome blocks. Thus, it may be necessary to clone and sequence QTL in order to identify the numbers of genes involved and their direct effects.

FIXING HIGH PARENT HETEROSIS IN SELF POLLINATED CROPS

Heterosis in self-pollinated crops also was discussed at the Heterosis Conference in 1952. Smith (1952) referred to high parent heterosis as transgressive vigor and showed that by using inbreeding and selection it was possible to develop improved lines of tobacco (*Nicotiana tabacum* L.) which exceeded the high parent and sometimes even the F_1 in most characteristics. This strategy is perhaps the mainstay of the pedigree method in breeding self-pollinated crops.

Powers (1952) reinforced the recovery of inbred lines in tomato [*Lycopersicon lycopersicum* (L.) Karsten] and barley (*Hordeum vulgare* L.) that retained the advantages attributed to heterosis. He used marker genes and was very explicit about the importance of blocks of linked genes in the process. In fact, Powers used the markers in tomato to provide proof of recombination of genes to produce segregates with greater weight per locule in F_2 and backcross segregates than in the F_1. He noted that if heterosis associated with several markers were due solely to an interaction of the marker genes as in the overdominance hypothesis, then it would not be possible to obtain homozygous lines possessing the increases. Thus, linked genes in complementing chromosome blocks provide an explanation of heterosis and associated gene action in breeding systems in both cross- and self-pollinated crops.

CHROMOSOME BLOCKS ARE A UNIFYING CONCEPT FOR HETEROSIS IN ALLO- AND AUTOPOLYPLOIDS

Allopolyploids (disomic polyploids) have fixed heterozygosity in the two or more genomes they possess {wheat (*Triticum aestivum* L.), oats (*Avena sativa* L.), soybean [*Glycine max* (L.) Merr.], cotton (*Gossypium hirsutum* L.), tobacco, and others}. Hexaploid wheat is self-pollinated and homozygous within each of its three genomes, but has potential heterozygosity and genetic complementation among its three genomes (Mac Key, 1970). Proof of this complementation is the famous nullisomic–tetrasomic compensation series in wheat (Sears 1966). Removal of a chromosome from one genome (nullisomic), can be compensated by extra homeologues (tetrasomic) in another genome. Homozygosity of an unfavorable allele in one genome may be complemented by a more favorable allele on a homeologue in another genome. Some of the nullisomic tetrasomic lines have a distinctive phenotype, but they are viable because of genomic compensation (Sears, 1966). Thus, disomic polyploids can benefit from the advantages of fixed hybridity and self-fertilization (Mac Key, 1970).

Autopolyploids (polysomic polyploids) ensure their heterozygosity through cross-pollination (potato, alfalfa, birdsfoot trefoil (*Lotus corniculatus* L.), and many forage grasses). Polysomic segregation of heterozygotes is known to protect recessive alleles from segregation in the homozygous condition. The familiar monohybrid (Aa) disomic F_2 segregation, where 1/4 homozygous recessives are revealed is reduced to a mere 1/36 in the equivalent tetrasomic segregation (AAaa). Segregations with two loci are even more extreme: the diploid is 9:3:3:1 versus the autotetraploid that is 1225:35:35:1.

The extent to which disomic and polysomic polyploids share a dependency on heterozygosity among genomes and at individual loci, respectively, was discussed previously (Mac Key, 1970; Bingham, 1980). These previous discussions focused on the polysomic locus, e.g., the tetrasomic locus with potentially four different alleles and allelic interactions (overdominance); however as noted earlier, alfalfa research has demonstrated that accumulation of favorable dominant alleles, not overdominance, explained genetic gains (Pfeiffer & Bingham 1982; Woodfield & Bingham, 1995). A genetic model involving complementary genes in repulsion phase linkage blocks explained results. As established earlier in maize, overdominance effects were due to complementary genes associated with linkage disequilibrium (Gardner & Lonnquist, 1959; Robinson et al., 1960; Moll & Robinson, 1967; Lonnquist, 1980).

A simple model involving two loci in repulsion phase linkage demonstrates the genetic similarity of allo- and autopolyploids due to chromosome blocks (Fig. 6–6a and 6–6b, repsectively). The model indicates that complete complementation (100%) in the disomic tetraploid (Fig. 6–6c), would still be 94% in the progeny of the tetrasomic tetraploid (Fig. 6–6d). The potential frequency of loci with dominants in the two types of polyploids is essentially similar. This is striking, especially considering that disomic polyploids are predominantly self-pollinated and that the polysomics are outcrossed. In spite of the different modes of reproduction, the two types of polyploids have similar genetic architecture. Thus chromosome blocks provide an explanation of the fixed heterozygosity in disomic

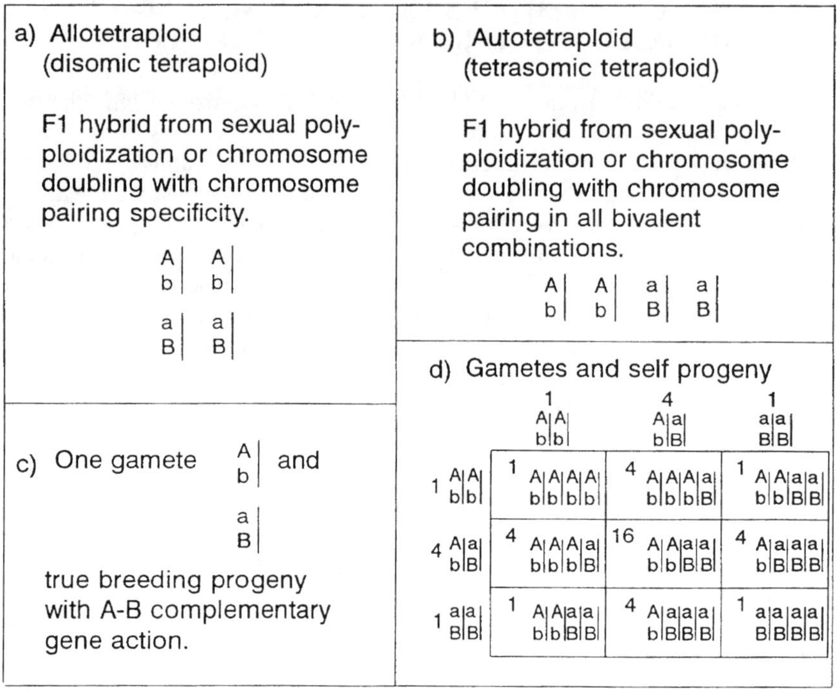

a) Allotetraploid (disomic tetraploid) F1 hybrid from sexual polyploidization or chromosome doubling with chromosome pairing specificity. $\begin{array}{cc} A & A \\ b & b \\ a & a \\ B & B \end{array}$	b) Autotetraploid (tetrasomic tetraploid) F1 hybrid from sexual polyploidization or chromosome doubling with chromosome pairing in all bivalent combinations. $\begin{array}{cccc} A & A & a & a \\ b & b & B & B \end{array}$
c) One gamete $\begin{array}{c} A \\ b \end{array}$ and $\begin{array}{c} a \\ B \end{array}$ true breeding progeny with A-B complementary gene action.	d) Gametes and self progeny

d) Gametes and self progeny:

	1 A|A| b|b|	4 A|a| b|B|	1 a|a| B|B|
1 A|A| b|b|	1 A|A|A|A| b|b|b|b|	4 A|A|A|a| b|b|b|B|	1 A|A|a|a| b|b|B|B|
4 A|a| b|B|	4 A|A|A|a| b|b|b|B|	16 A|A|a|a| b|b|B|B|	4 A|a|a|a| b|B|B|B|
1 a|a| B|B|	1 A|A|a|a| b|b|B|B|	4 A|a|a|a| b|B|B|B|	1 a|a|a|a| B|B|B|B|

Fig. 6–6. Theoretical model of a chromosome block with two complementary loci, each with a dominant allele linked in repulsion. The consequences of using the chromosome block to produce an allotetraploid (disomic tetraploid) and an autotetraploid (tetrasomic tetraploid) are illustrated. See text for discussion.

polyploids, the high level of heterozygosity in polysomic polyploids, and the pseudo-overdominance in diploid hybrids.

SUMMARY AND CONCLUSIONS

Chromosome blocks are the unit of genetic transmission that must be considered in genetic models to explain heterosis. Individual genes are linked in chromosome blocks. Thus, estimates of gene action are due to the cumulative effect of linked blocks of genes. The genes in chromosome blocks are in linkage disequilibrium in early generations after crossing, and approach linkage equilibrium after about eight generations of random mating. Even then it is the population that is approaching linkage equilibrium; thus an individual in the population that is withdrawn and used in a cross resets the linkage disequilibrium clock.

All who have written about overdominance versus pseudo-overdominance due to linkage disequilibrium agree that the possibility of true overdominance at some loci may well occur; however, the largest component of overdominance estimates must be due to pseudo-overdominance, according to mounting evidence. Action of single genes can only be studied as changes at the locus due to new mutation, transposable elements, single gene transformations, or by kinetic data on known single gene products. This restricts the scope of potential research on true overdominance.

Improvement of inbred lines in essentially all long-term plant breeding programs must be due to accumulation of favorable genes in chromosome blocks and their associated interactions. The same can be said for population improvement of metric traits. This is the strength of plant breeding and reason for optimism about continued plant improvement.

Chromosome blocks provide an efficient way of masking several deleterious recessive alleles at once. It is concluded that linked genes on chromosome blocks provide an explanation of heterosis in diploids, the fixed heterosis in self-pollinated allopolyploids, and the relatively high levels of heterosis maintained under cross-pollination in autopolyploids. Thus, chromosome blocks provide a unifying concept for all categories of plants.

REFERENCES

Bingham, E.T. 1980. Maximizing heterozygosity in autopolyploids. p. 471–489. *In* W.H. Lewis (ed.) Polyploidy. Plenum Press, New York.

Bingham, E.T., R.W. Groose, D.R. Woodfield, and K.K. Kidwell. 1994. Complementary gene interactions in alfalfa are greater in autotetraploids than diploids. Crop Sci. 34:823–829.

Bruce, A.B. 1910. The Mendelian theory of heredity and the augmentation of vigor. Science (Washington, DC) 32:627–628.

Burton, G.W. 1986. Identifying heterotic blocks on pearl millet chromosomes with chlorophyll deficient mutants. Crop Sci. 26:537–539.

Burton, G.W., and B.K. Werner. 1991. Genetic markers to locate and transfer heterotic chromosome blocks for increased pearl millet yields. Crop Sci. 31:576–579.

Burton, G.W., and J.P. Wilson. 1995. Identification and transfer of heterotic chromosome blocks for forage yield in pearl millet land races. Crop Sci. 35:1184–1187.

Brewbaker, J.L. 1964. Agricultural genetics. Prentice-Hall, Englewood Cliffs, NJ.

Cockerham, C.C., and Z-B Zeng. 1996. Design III with marker loci. Genetics 143:1437–1456.

Collins, G.N. 1921. Dominance and the vigor of first generation hybrids. Am Nat. 55:116–133.

Comstock, R.E., and H.F. Robinson. 1952. Estimation of average dominance of genes. p. 494–517. *In* J.W. Gowen (ed) Heterosis. Iowa State College Press, Ames.

Crow, J.F. 1952. Dominance and overdominance. p. 282–297. *In* J.W. Gowen (ed.) Heterosis. Iowa State College Press, Ames.

Crow, J.F. 1993. Mutation, mean fitness, and genetic load. Oxford Surv. Evol. Biol. 9:3–42.

Davenport, C.B. 1908. Degeneration, albinism and inbreeding. Science (Washington, DC) 28:454–455.

Demarly, Y. 1963. Genetique des tetraploids et amelioration des plantes. Ann. Amelio. Plant. 13:307–400.

Demarly, Y. 1979. The concept of linkat. p. 257–265. *In* A.C. Zeven and A.M. van Harten (ed.) Proc. Conf. Broadening Genetic Base of Crops, Wageningen, the Netherlands. 3–7 July 1978. Ctr. for Agric. Publ. and Doc., Wageningen, the Netherlands.

Dobzhansky, 1952. Nature and origin of heterosis. p. 218–224. *In* J.W. Gowen (ed.) Heterosis. Iowa State College Press, Ames.

Doebley, J., A. Stee, and C. Gustus. 1995. Teosinte branched 1 and the origin of maize: Evidence for epistasis and the evolution of dominance. Genetics 141:333–346.

Dudley, J.W., R.J. Lambert, and D.E. Alexander. 1974. Seventy generations of selection for oil and protein concentration in the maize kernel. *In* J.W. Dudley (ed.) Seventy generations of selection in maize. ASA, Madison, WI.

Dunbier, M.W., and E.T. Bingham. 1975. Maximum heterozygosity in alfalfa: Results using haploid-derived autotetraploids. Crop Sci. 15:527–531.

East, E.M. 1908. Inbreeding in corn. Rep. Connecticut Agric. Exp. Stn. 1907:419–429.

East, E.M., and H.K. Hayes. 1912. Heterozygosis in evolution and plant breeding. USDA Bureau of Plant Industry Bull. 243. USDA, Washington, DC.

Emerson, R.A., and E.M. East. 1913. The inheritance of quantitative characters in maize. Nebraska Agric. Exp. Stn. Bull. 2.

Fisher, R.A. 1918. The correlation between relatives on the supposition of Mendelian inheritance. Trans. Royal Soc. Edinburgh. 52:399–433.

Fisher, R.A. 1931. The evolution of dominance. Biol. Rev. Cambridge Phil. Soc. 6:345–368.

Gardner, C.O., P.H. Harvey, R.E. Comstock, and H.F. Robinson. 1953. Dominance of genes controlling quantitative characters in maize. Agron. J. 45:186–191.

Gardner, C.O., and J.H. Lonnquist. 1959. Linkage and the degree of dominance of genes controlling quantitative characters in maize. Agron. J. 51:524–528.

Gowen, J.W. 1952. Heterosis. Iowa State College Press, Ames.

Groose, R.W., L.E. Talbert, W.P. Kojis, and E.T. Bingham. 1989. Progressive heterosis in autotetraploid alfalfa: Studies using two types of inbreds. Crop Sci. 29:1173–1177.

Hallauer, A.R., and J.B. Miranda. 1988. Quantitative genetics in maize breeding. 2nd ed. Iowa State Univ. Press, Ames.

Hull, F.H. 1952. Recurrent selection for overdominance. p. 451–474. *In* J.W. Gowen (ed.) Heterosis. Iowa State College Press, Ames.

Jinks, J.L. 1983. Biometrical genetics of heterosis. p. 1–46. *In* R. Frankel (ed.) Heterosis: A reappraisal of theory and practice. Springer-Verlag, New York.

Jones, D.F. 1917. Dominance of linked factors as a means of accounting for heterosis. Genetics 2:466–479.

Jones, D.F. 1952. Plasmagenes and chromogenes in heterosis. p. 224–225. *In* J.W. Gowen (ed.) Heterosis. Iowa State College Press, Ames.

Keeble, F., and Pellew, C. 1910. The mode of inheritance of stature and of time of flowering in peas (*Pisum sativum*), J. Genet. 1:47–56.

Khavkin, E.E., and E.H. Coe. 1995a. Organization of growth-regulating genes in maize: 1. The functional clusters of genes. Russ. J. Plant Phys. 42:408–420.

Khavkin, E.E., and E.H. Coe. 1995b. The organization of growth-regulating genes in maize: 2. Quantitative trait loci. Russ. J. Plant Phys. 42:558–574.

Lamkey, K.R., A.R. Hallaner, and D.S. Robertson. 1988. Contribution of the long arm of chromosome 10 to the total heterosis observed in five maize hybrids. Crop Sci. 28:896–902.

Lonnquist, J.H. 1980. Heterosis: Additivity versus dominance. An. Acad. Nac. Cienc. Ex. Fis. Nat. (Buenos Aires) 32:195–202.

Mac Key, J. 1970. Significance of mating systems for chromosomes and gametes. Hereditas 66:165–176.

Mangelsdorf, P.C. 1952. Hybridization in the evolution of maize. *In* J.W. Gowen (ed.) Heterosis. Iowa State College Press, Ames.

Mather, K., and J.L. Jinks. 1973. Introduction to biometrical genetics. Cornell Univ. Press, Ithaca, NY.

Moll, R.H., and H.F. Robinson. 1967. Quantitative genetic investigation of yield of maize. Der Zuchter 37:192–199.

Peterson, P.A., and E. Wernsman. 1964. A monosomic-type approach to maize breeding. Crop Sci. 4:533–535.

Pfeiffer, T.W., and E.T. Bingham. 1983. Improvement of fertility and herbage yield by selection within two-allele populations of alfalfa. Crop Sci. 23:633–636.

Powers, L. 1952. Gene recombination and heterosis. *In* J.W. Gowen (ed.) Heterosis. Iowa State College Press, Ames.

Rhoades, D., E.C. Ju, W-J Yang, and Y. Samaras. 1992. Plant metabolism and heterosis. Plant Breed. Rev. 10:54–89.

Robertson, D. 1964. Transfer of intact segments of maize chromosomes: A possible method. J. Heredity 55:107–114.

Robinson, H.F., C.C. Cockerham, and R.H. Moll. 1960. Studies on estimation of dominance variance and effects of linkage bias. p. 171–177. *In* Biometrical genetics. Int. Ser. of Monogr. on Biometry. Pergamon Press, New York.

Robinson, H.F., R.E. Comstock, and P.H. Harvey. 1949. Estimates of heritability and the degree of dominance in corn. Agron. J. 41:353–359.

Robinson, H.F., R.E. Comstock, and P.H. Harvey. 1955. Genetic variances in open pollinated varieties of corn. Genetics 40:45–60.

Sears, E.R. 1966. Nullisomic-tetrasomic combinations in hexaploid wheat. *In* R. Riley and R.R. Lewis (ed.) Chromosome manipulations in plants. Oliver & Boyd, London.

Shull, G.H. 1908. The composition of a field of maize. Rep. Am. Breed. Assoc. 4:296–301.
Shull, G.H. 1911. The genotypes of maize. Am Nat. 45:234–252.
Shull, G.H. 1952. Beginnings of the heterosis concept. p. 14–48. *In* J.W. Gowen (ed.) Heterosis. Iowa State College Press, Ames.
Smith, H.H. 1952. Fixing transgressive vigor in *Nicotiana rustica*. p. 161–174. *In* J.W. Gowen (ed.) Heterosis. Iowa State College Press, Ames.
Sprague, G.F. 1983. Heterosis in maize; theory and practice. p. 47–70. *In* R. Frankel (ed.) Heterosis: A reappraisal of theory and practice. Springer-Verlag, New York.
Stuber, C.W. 1992. Biochemical and molecular markers in plant breeding. Plant Breed. Rev. 9:37–57.
Stuber, C.W. 1994. Heterosis in plant breeding. Plant Breed. Rev. 12:227–251.
Stuber, C.W., S.E. Lincoln, D.W. Wolff, T. Helentjaris, and E.S. Lander. 1992. Identification of genetic factors contributing to heterosis in a hybrid from two elite maize inbred lines using molecular markers. Genetics 132:823–839.
Woodfield, D.R., and E.T. Bingham. 1995. Improvement in two-allele autotetraploid populations of alfalfa explained by accumulation of favorable alleles. Crop Sci. 35:988–994.

7 Inference of Heterosis and Epistasis in Transposon Tagged *Drosophila*

Andrew G. Clark

Institute of Molecular Evolutionary Genetics
Pennsylvania State University
University Park, Pennsylvania

ABSTRACT

One of the problems associated with statistical inference of gene interactions has been the relatively low power of most methods. By constructing known genotypes and measuring the phenotypes of these genotypes, novel and powerful tests of effects of mutations on phenotypic expression can be performed. Clark and others produced and analyzed a set of 263 lines of *Drosophila melanogaster* that had single, random transposable *P*-element insertions in the genome. Activities of 12 enzymes in intermediary metabolism were quantified along with levels of stored triacylglycerol and glycogen, total protein, and live weight. The tagged lines exhibited significant heterogeneity in 11 of the 16 metabolic characters, and a surprising 58% of the *P*-element insertions affected at least one character. This result implies that these genes are regulated through a highly interactive network, such that many changes in the genome affect the expression of many genes. Simple crosses among such *P*-element tagged lines allow for the construction of all possible two-locus genotypes for each pair of *P*-elements. Analysis of phenotypes of the nine genotypes provides direct estimates of additive, dominance, and epistatic effects of the mutations. Least squares fits to linear models of Cockerham indicate that 14% of the tests exhibited significant epistatic effects, and of the significant tests, 74% are either additive × additive or dominance × dominance epistasis. In general, it appears that novel mutations have epistatic effects on metabolic traits that are on the same order of magnitude as main (additive and dominance) effects. *P*-element insertions whose effects were heterotic also were significantly more likely to be those that manifested dominance × dominance epistasis.

To most geneticists, dominance and epistasis are viewed as properties intrinsic to a gene. Quantitative geneticists think in terms of partitioning variance in a population, so levels of dominance and epistasis for a trait vary along with gene frequency. But concepts like epistasis and heterosis also have meaning apart from the variance partitioning approach. Considering the nine possible two-locus genotypes, it is clear that the phenotypes associated with each of these genotypes may or may not fall in a pattern that is additive, regardless of the population fre-

---

Copyright © 1998 Crop Science Society of America, 677 S. Segoe Rd., Madison, WI 53711, USA.
Concepts and Breeding of Heterosis in Crop Plants. CSSA Special Publication no. 25.

quencies. Knowledge of genotypes can be gained by either scoring molecular markers or by constructing genotypes by genetic crosses. By directly scoring the phenotypes of these nine genotypes, models can be fitted that allow precise quantification of the magnitude and types of epistasis. One well-respected model is that of Cockerham (1954), which partitions the eight degrees of freedom in such an experiment into two additive terms, two dominance terms, and four epistatic terms: additive × additive, additive × dominance, dominance × additive, and dominance × dominance.

Measured genotypes provide a powerful approach for resolving and quantifying epistatic interactions. Classical quantitative genetic methods for inferring epistasis have long suffered from a lack of statistical power in the sense that the power to detect additive and dominance components has always been greater than the power to detect epistasis. Measured genotypes allow tests of epistasis that nearly match the statistical power of tests of additivity and dominance. A classic study of epistasis in identified genomic segments entailed construction of all nine two-locus genotypes of isogenic lines of barley (Fasoulas & Allard, 1962). Measurements of seven morphological characters in these plants revealed that 32% of the genotypic variance was epistatic. Russell (1976) took the analysis to three loci, and after constructing and testing all 27 genotypes, he found 47% of all tests of additive × additive epistasis significant at the 5% level. These results suggested that a controlled genetic background may greatly improve the sensitivity to detect epistasis and with this improvement comes the realization that epistasis may be common.

Other studies of crop plants generally employ a more heterogeneous genetic background and test for marginal epistasis with respect to two of the many segregating factors. In the case of maize (*Zea mays* L.), despite the increased power, relatively little epistasis was found even by measured-genotype methods. Edwards et al. (1987) examined two sets of F_2 maize populations by scoring 20 molecular markers and 82 morphological traits. Consistent with classical estimates of genetic variance components in maize, they found relatively little digenic epistasis (~3% of interaction terms in a two-way ANOVA were significant). Measured genotype methods in fact have revealed relatively little epistasis in any plant species (Tanksley, 1993), except in some wide crosses, such as between maize and teosinte (Doebley & Stec, 1991; Doebley et al., 1995). In *Drosophila*, one does not have to look at wide crosses to find pronounced levels of epistasis. Long et al. (1995) tested F_2 flies derived from crosses of high- and low-bristle number selected lines. Of the 60 tests that could be performed, 13 detected significant epistatic effects. Moreover, the magnitude of epistatic effects was similar to the magnitude of additive effects. The list of tests of epistasis with measured genotypes is growing rapidly, and a surprising amount of epistasis is being revealed.

Measured genotypes also may provide important insights into the cause of heterosis. By quantifying heterotic effects associated with molecular markers, it is possible to map portions of the genome responsible for heterosis. One potential cause of heterosis is epistasis. If both parental types are inbreds with poor combinations of genes, the F_1 may possess gene combinations that provide superior fitness. In this study, the transposable *P*-element is used to quantify both epistasis and heterosis in the same lines in order to directly measure the association. The effects of pairs of *P*-element insertions on 16 quantitative characters are quantified in order to determine relative magnitudes of additive, dominance, and epistatic effects and the degree of heterosis in 128 independent tests. Simultaneous evaluation of these tests allows assessment of patterns of significant epistasis across genotypes and across traits.

MATERIALS AND METHODS

Drosophila Stocks

Lines of *Drosophila* bearing single *P*-elements in a homogeneous genetic background were constructed by mobilizing the CaSpeR transposable element with the $\Delta 2$-3 *P*-element as a transposase source (Cooley et al., 1988; Robertson et al., 1988). Details of the initial stock construction appear in Clark et al. (1995). Sixteen lines were selected from the original set of 263 *P*-element tagged lines based on their stable low level expression of the *white* eye pigmentation gene. For each of eight pairs of *P*-element insertion lines, a second-chromosome and a third-chromosome insertion line were crossed to produce F_1 flies, which were then crossed to give an F_2 generation that exhibited a wide range in eye color phenotypes. The line with the darkest pigmentation was putatively the line doubly homozygous for the *P*-element insertions. Virgin stocks were crossed to produce a stable double insert strain (*aabb*), and its genotype was verified genetically. Crosses between the single insert strains (*aaBB* and *AAbb*), the double insert strain, and the noinsert *white* eye strain (*AABB*) were used to produce nine two-locus genotypes. The doubly heterozygous line was made both by crossing the two single insert homozygotes and by crossing the double insert line with the zero insert line. These two crosses should produce identical genotypes in the offspring, and both were tested for all quantitative phenotypes. Each genotype was constructed by two independent sets of crosses, and analysis of variance was used to test for homogeneity in phenotypes.

Assays of Enzyme Activities

Drosophila stocks were reared under uniform conditions, anesthetized, weighed, homogenized, centrifuged, and distributed into microtiter plates as described previously (Clark et al., 1995). For each sample of flies, 16 characters were quantified, including live weight (WT), and 15 biochemical quantities that were assayed with a microtiter plate reader (VMAX, Molecular Devices, Menlo Park, CA). Further details of the procedures appeared in Clark and Wang (1994). Each microtiter plate contained homogenate samples as well as standards and controls.

The 15 identical plates were used to assay triacylglycerol content (TRI), glycogen content (GLY), total protein content (PRO), and the activities of alcohol dehydrogenase (ADH, at map location 2-50.1), fatty acid synthase (FAS), glucose6phosphate dehydrogenase (G6PD, map location 1-63.0), α-glycerol-3-

phosphate dehydrogenase (GPDH, map location 3-55.4), glycogen phosphorylase (GP), glycogen synthase (GS), hexokinase (HEX, Hex-A at 1-29.3 and Hex-C at 2-75.0), malic enzyme, (ME, at 3-53.1), 6-phosphogluconate dehydrogenase (6PGD, at 1-0.64), phosphoglucose isomerase (PGI, at 2-58.0), phosphoglucomutase (PGM, at 3-43.4), and trehalase (TRE, at 2-92.0). The units of TRI and GLY are reported as micrograms per fly, and the units of the enzyme activities are given as nM NADP (or NAD) reduced per fly per minute. The microtiter plate reader was programmed to record 6 to 12 optical densities of each well at prescribed time intervals depending on the assay. The design included 8 P-element pairs × 10 genotypes per pair × 2 replicate constructions × 4 replicate assays.

Statistical Analysis

The data were fitted to linear models that had the classifications: second chromosome genotype (A_i, where i = 0, 1, or 2 copies of the P-element), third chromosome genotype (B_j, where j = 0,1,2), replicate construct (C_{ijk}), and an error term (e_{ijkl}) representing differences among replicate samples. Null hypotheses of interest included the equivalence of the two doubly heterozygous lines, additivity of effects of the two P-element inserts, and fits to models that parameterize epistasis. These null hypotheses were tested separately for all 16 characters assayed and for each of the eight P-element pairs. The following model for analysis of variance was used for the simple test of additivity of the two P-elements' effects:

$$Y_{ijkl} = \mu + A_i + B_j + (A \times B)_{ij} + C_{ijk} + e_{ijkl}$$

where μ is the grand mean, and $(AB)_{ij}$ is the term representing the interaction between effects of the second and third chromosome P-element insertion.

The 14 traits other than live weight and total protein content are expected to be correlated with body size, so the covariates weight and protein were removed by performing analysis of covariance, as described by Clark and Wang (1997). Models that explicitly incorporate parameters of epistasis were fitted by regression as described by Clark and Wang (1997). Heterosis was inferred by posterior contrasts in the linear models testing whether heterozygotes for each P-element insert had a phenotype outside the range of the two homozygous genotypes. Associations between epistasis and heterosis were examined by simple chisquare tests on counts of cases with and without heterosis and epistasis, under the assumption that each trial was independent.

RESULTS

Clark and Wang (1997) showed that independent constructs of each genotype were not significantly different in the phenotypes. This was important to demonstrate because there is the possibility that P-element transpositions can introduce mutations at sites other than the insertion site. Apparently such mutations were sufficiently rare, and/or their effects were sufficiently small that they can be ignored. Clark and Wang (1997) also verified that the double heterozygous lines derived from a cross of the two double homozygotes ($aabb \times AABB$) had the same phenotype as double heterozygous lines obtained by crossing the two singleinsert homozygotes ($aaBB \times AAbb$).

Simple Two-Way ANOVA

Phenotype scores were examined by analysis of variance, treating each gene as a fixed effect and examining whether the genotypes at the two loci act nonadditively. The standard test for significance of the interaction term in ANOVA provides one measure of overall epistasis. Of the 128 tests for an epistatic interaction between the two P-elements (8 P-element pairs × 16 traits), 29 were found to be significant at the 5% level after correcting for multiple comparisons (Clark & Wang, 1997). This relatively high figure suggests that a closer inspection of the structure of the epistatic effects was warranted. Analysis of covariance, correcting for the covariates live weight and total protein, yielded 31 significant interaction terms. For quantitative genetic analysis it is important to choose an appropriate scaling of the data, because a trait that appears to be dominant on a linear scale may be additive on a log scale. When the above ANOVA was repeated on a logarithmic scale, nearly indistinguishable results were obtained (33 of the 128 tests were significant at $P < 0.05$). The magnitude of epistasis revealed in the following regression analyses was likewise robust with respect to scaling.

Regression Estimates of Variance Component

One of the most widely used parameterizations of epistasis was developed by Cockerham (1954). He showed that the nine phenotypes for two diallelic loci could be tested in eight orthogonal contrasts, corresponding to the two additive effects parameters, two dominance parameters, and four epistatic terms, additive × additive (A × A), additive × dominance (A × D), dominance × additive (D × A), and dominance × dominance (D × D).

The epistatic parameters in this study were fitted by regressing each phenotype on corresponding index variables for each genotype (Cockerham, 1954). A parameter estimate was judged to be significantly different from zero if it departed by more than two sample standard errors from zero. It was assumed that the lines were in Hardy-Weinberg equilibrium with allele frequency of ½. A striking result was that the epistatic variance components were as large in magnitude as the first order terms for levels of additive and dominance variance (Clark & Wang, 1997). Results of the significance tests appear in Fig. 7–1, which shows all 512 tests (8 P-element pairs × 16 traits × 4 epistasis parameters). The heterogeneity across P-element pairs is readily apparent (e.g., Set 3 exhibited very little epistasis, while Set 2 exhibited considerable interaction). Moreover, if one trait exhibited particularly strong A × D epistasis, then other traits were more likely to show the same component as significant (e.g., Set 5).

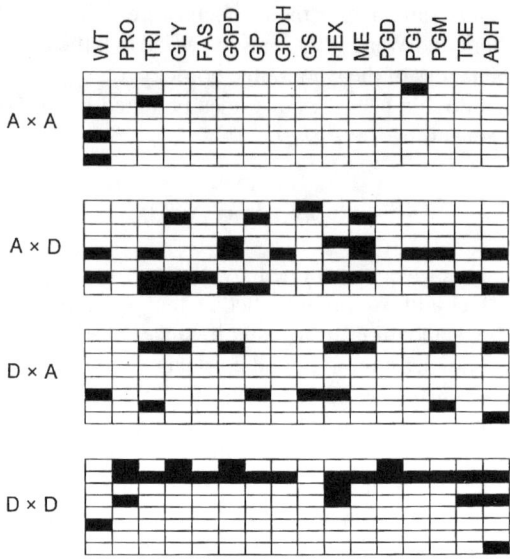

Fig. 7–1. Epistatic parameters for additive × additive (A × A), additive × dominance (A × D), dominance × additive (D × A) and dominance × dominance (D × D) epistasis from Cockerham's (1954) model. Tests that deviated significantly from zero (at $P < 0.05$) are plotted as black rectangles. Each block shows one of the four epistatic parameters tested for all eight P-element pairs × 16 traits.

Heterosis and Epistasis

Each of the 16 traits for each of the eight P-element pairs (128 tests) were scored for significant epistatic components as described above. The marginal phenotypic means were calculated for a population with allele frequency ½, and comparative analyses were done to determine whether the heterozygote had a phenotype greater than the higher homozygote. If the issue of statistical significance is ignored, several cases of heterosis were seen (Fig. 7–2). Figures 7–1 and 7–2 were compared to determine whether the cases that exhibit significant epistasis are also more likely to exhibit heterosis. Comparisons were made by chi-

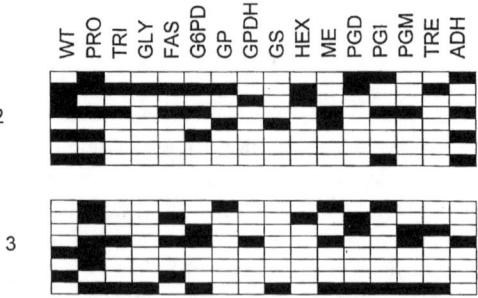

Fig. 7–2. Combinations of lines and traits exhibiting heterosis, defined as having a mean heterozygous phenotype falling outside the range of the two homozygotes (not necessarily statistically significant). The top block is for the P-element insertions on the second chromosome, and the bottom block is for third chromosome insertions.

Table 7–1. Contingency tables for tests of association between epistasis and heterosis.

		Dominance × Dominance epistasis		
		No	Yes	Row total
Heterosis in either P-insert	No	82†	9	91
	Yes	21	16	37
	Column total	103	25	128
		$X_1^2 = 18.619$	$P < 0.0001$	
		Dominance × Dominance epistasis		
		No	Yes	Row total
Heterosis in both P-inserts	No	94	18	112
	Yes	9	7	16
	Column total	103	25	128
		$X_1^2 = 6.824$	$P < 0.009$	

† Counts of tests of heterosis (phenotype outside the range of both homozygotes) and of dominance (dominance epistasis for the 128 combinations of P-element insertions and 16 traits were tallied.

square contingency tests for each component of epistasis, and only the D × D component was significant (Table 7–1). In such cases, lines and traits with D × D epistasis were more likely to exhibit heterosis. Heterosis enters the model through the dominance terms, and it is statistically orthogonal to D × D epistasis. This means that the significant association in Table 7–1 is not an artifact of the parameterization, but may have biological meaning. These results should be interpreted with some caution, however, because the experiments were not designed to have much statistical power for detecting heterosis. Application of linear models showed that only two of the cases (protein levels in P-element Set 1 and live weight in P-element Set 7) were significant.

DISCUSSION

The data and analysis presented here allow several firm conclusions to be drawn. First, many cases of significant epistasis were detected. Twenty-three percent of the analyses of variance exhibited a significant interaction term, and models that specifically parameterize components of epistasis revealed that 15% of the tests gave significant epistatic components. Second, there was heterogeneity among P-element pairs in the extent of epistasis exhibited. Some lines exhibited extensive epistasis in several traits, while others appeared nearly free of interaction. Third, the additive × dominance epistasis was directional in the sense that if A × D was significant for a given trait and P-element pair, other traits for that P-element pair tended to have A × D and not D × A epistasis. Fourth, the distribution of significant components of Cockerham's tests was not random. That is, if a particular component was significant for one trait, it is more likely to be significant for other traits as well. This was caused in part by correlations across traits due to pleiotropic effects of the P-element insertions (Clark et al., 1995). Finally, there was a significant tendency for trait × P-element combinations to show associations between heterosis and epistasis. That is, heterosis in marginal effects (collapsing over genotypes at one or the other P-element insertion site), was more likely to be seen if the P-element pair exhibited significant epistasis in that trait.

Both heterosis and epistasis tend to be particularly prevalent in crosses among most widely divergent races and among interspecific crosses (Whitlock et al., 1995). Particularly good examples are seen in the genes that cause interspecific hybrid male sterility in the *D. melanogaster* subgroup (Cabot et al., 1994; Palopoli & Wu, 1994). This may be because polymorphism for strong interactions are not retained in populations, which may be due to their frequent lack of global stability. Alternatively, wide crosses may result in stronger effects at all levels (including pleiotropic effects, which ultimately result in apparent interactions) even if the interactions are very indirect. In either case, the consequences for gene dynamics and response to selection are real and need to be accommodated in population genetics models.

Some traits are intrinsically more likely to exhibit epistatic interactions among underlying genes than are other traits. Because there are biochemical interactions among components of the transcription complex, it is possible that the determinants of an individual gene's expression may be highly nonadditive. The assembly of a transcription complex occurs only through delicate pairwise and multiway interactions among a large set of transcription factors and related proteins. From this molecular viewpoint, genetic variation in the factors that result in that gene's expression are likely to interact. Thus, traits that are likely to be manifestations of expression of one or a few genes are likely to be epistatic. Our results, showing extensive epistasis among P-element inserts that affect enzyme activities, supporting this hypothesis. Similarly, Damerval et al. (1994) examined expression of 72 proteins in 85 F_2 maize plants from a distant cross and found that 14% of the proteins exhibited significant epistasis between genomic regions. Historically, examination of morphological traits in maize have repeatedly exhibited little epistasis, however, traits close to the expression of individual genes appear to exhibit significant levels of epistasis.

This study examined epistatic and heterotic effects of new mutations caused by P-element insertion. Such effects are likely to be different from the distribution of epistatic effects in an equilibrium population. Our tests were performed in an isogenic background, and just as in the study of Fasoulas and Allard (1962), epistatic interactions were abundant. Natural populations are segregating at many other loci, and the heterogeneous genetic background may obscure particular pairwise interactions. Large differences between levels of mutational epistasis and equilibrium epistasis may provide evidence for the action of natural selection removing a nonrandom set of alleles from the population. A test for this kind of selection could be done after quantifying levels of epistasis in a natural population. Measured genotype methods provide new opportunities to disentangle components of epistasis with a resolving power that will even allow tests of mechanisms of gene interactions.

ACKNOWLEDGMENTS

I thank Lei Wang and James Rosinski for technical assistance. This work was supported by National Science Foundation grant DEB9419631.

REFERENCES

Cabot, E.L., A.W. Davis, N.A. Johnson, and C.I. Wu. 1994. Genetics of reproductive isolation in the *D. simulans* clade: Complex epistasis underlying hybrid male sterility. Genetics 137:175–189.

Clark, A.G., and L. Wang. 1994. Comparative evolutionary analysis of metabolism in nine *Drosophila* species. Evolution 48:1230–1243.

Clark, A.G., and L. Wang. 1997. Epistasis in measured genotypes: *Drosophila* P-element insertions. Genetics 147:157–163.

Clark, A.G., L. Wang, and T. Hulleberg. 1995. P-element induced variation in metabolic regulation in *Drosophila*. Genetics 139:337–348.

Cockerham, C.C. 1954. An extension of the concept of partitioning hereditary variance for analysis of covariances among relatives when epistasis is present. Genetics 39:859–882.

Cooley, L., R. Kelley, and A. Spradling. 1988. Insertional mutagenesis of the *Drosophila* genome with single P elements. Science (Washington, DC) 239:1121–1128.

Damerval, C., A. Maurice, J.M. Josse, and D. de Vienne. 1994. Quantitative trait loci underlying gene product variation: A novel perspective for analyzing regulation of genome expression. Genetics 137:289–301.

Doebley, J., and A. Stec. 1991. Genetic analysis of the morphological differences between maize and teosinte. Genetics 129:285–295.

Doebley, J., A. Stec, and C. Gustus. 1995. Teosinte branched 1 and the origin of maize: Evidence for epistasis and the evolution of dominance. Genetics 141:333–346.

Edwards, M.D., C.W. Stuber, and J.F. Wendel. 1987. Molecular marker facilitated investigations of quantitative trait loci in maize: I. Numbers, genomic distribution and types of gene action. Genetics 116:113–125.

Fasoulas, A.C., and R.W. Allard. 1962. Nonallelic gene interactions in the inheritance of quantitative characters in barley. Genetics 47:899–907.

Long, A.D., S.L. Mullaney, L.A. Reid, J.D. Fry, C.H. Langley, and T.F.C. Mackay. 1995. High resolution mapping of genetic factors affecting abdominal bristle number in *D. melanogaster*. Genetics 139:1273–1291.

Palopoli, M.F., and C.I. Wu. 1994. Genetics of hybrid male sterility between *Drosophila* sibling species: A complex web of epistasis is revealed in interspecific studies. Genetics 138:329–341.

Robertson, H.M., C.R. Preston, R.W. Phillis, D.M. Johnson-Schlitz, W.K. Benz, and W.R. Engels. 1988. A stable genomic source of P-element transposase in *Drosophila melanogaster*. Genetics 118:461–470.

Russell, W.R. 1976. Genetic effects and genetic effect × year interactions at three gene loci in sublines of a maize inbred line. Can. J. Genet. Cytol. 18:23–33.

Tanksley, D.S. 1993. Mapping polygenes. Annu. Rev. Genet. 27:205–233.

Whitlock, M.C., P.C. Phillips, F.B.G. Moore, and S.J. Tonsor. 1995. Multiple fitness peaks and epistasis. Annu. Rev. Ecol. Syst. 26:601–629.

8 Avoiding Project Bankruptcy While Effectively Employing Markers

Tom Blake and Vladimir Kanazin

Department of Plant, Soil and Environmental Sciences
Montana State University
Bozeman, Montana

Steve Larson

USDA-ARS, Small Grains Germplasm Center
Aberdeen, Idaho

Joy Eckhoff

Eastern Agricultural Research Center
Montana Agricultural Experiment Station
Sidney, Montana

ABSTRACT

Most university plant breeding projects share similarities with small businesses. We have customers, our producers and funding agencies, and we have spheres of interest which are largely defined by the needs of our customers. Unlike our more basic research colleagues, *it's interesting* is not sufficient justification for a project. Our projects must also provide practical benefits to crop producers, while remaining sufficiently topical to attract grant funds and contribute to our field. Like a small business, we have enormous freedom to explore our working environment, and to find niches that are too small or specialized to be noticed by larger organizations. Many of these niches can be both lucrative to producers and fundamentally interesting. This chapter will discuss and describe how my project is addressing the needs of my primary customers, the small grains producers of Montana, while helping to develop the tools of barley (*Hordeum vulgare* L.) genetics for my secondary customers, research funding agencies.

GENERAL CHARACTERISTICS OF THE SMALL BREEDING PROJECT

The economy of Montana is built upon agriculture—primarily livestock, wheat (*Triticum aestivum* L.) and barley production. Our production system is an old one, built upon a crop–fallow rotation across much of the state. The inter-

---

Copyright © 1998 Crop Science Society of America, 677 S. Segoe Rd., Madison, WI 53711, USA. *Concepts and Breeding of Heterosis in Crop Plants.* CSSA Special Publication no. 25.

mountain valleys are enormously productive, although late season rains often damage grain quality. The eastern slope of the Rockies generally produces higher quality barley and wheat, with drought our greatest yield-limiting factor. Dark Northern Spring Wheat and 2-rowed malting barley are generally the most profitable crops in the state, while the high volume commodities produced in the state are hard red winter wheat and feed barley.

The job of the state barley breeder is to identify the most lucrative barley varieties that farmers can produce, and to develop varieties that further improve production profitability. Our research resources include a seven-farm research farm system staffed by competent agronomists, a molecular genetics laboratory, a state-of-the-art greenhouse facility, and a conservative but competent farming community consisting of a few thousand large land holders. Annually about 1.4 million acres are seeded to barley in Montana, with about 60% of the acreage seeded to varieties recommended as malting barley varieties by the American Malting Barley Association. Approximately 7 million acres are annually seeded to either winter or spring wheat.

PROBLEMS AND OPPORTUNITIES IN MONTANA

Although barley has a large and well-explored primary germplasm pool each of the germplasm groups currently in production represents a narrow and specialized subgroup (Martin et al., 1991; Hayes et al., 1997). The upper Midwest has historically produced 6-rowed malting barley used predominantly for domestic malt. Montana and Idaho have been the primary sources of 2-rowed malting barley in the USA. The reasons underlying the germplasm group differences are largely historical, although breeders have taken advantage of available genetic resources to create modified germplasm groups with excellent regional adaptation. The 2-rowed germplasm group descended largely from 'Betzes' and Betzes relatives shows excellent malting quality and good drought tolerance in the Pacific Northwest, while the Manchuria-derived 6-rowed germplasm group developed by Don Rasmussen produces excellent yields and superb quality grain in the upper Midwest. Unfortunately for Montana producers, the 'Manchuria'-derived varieties that perform so admirably in the Red River Valley suffer from the short, dry growing conditions that predominate throughout Montana. The challenge of adjusting the adaptational characteristics of a germplasm group to fit a substantially different production environment, while maintaining the quality characteristics that define the value of the group provides impetus to an applied geneticist.

The recent increase in scab infestation in the upper Midwest has proven to be a significant production problem for the USA and Canadian wheat and barley producers, and the introduction of a notably better-yielding feed barley variety from Germany ('Baronesse') has placed new emphasis on yield potential in the Pacific Northwest. These sorts of germplasm limitations generate the conditions needed to make genome research a productive addition to a small breeding program. When these events are combined with the current global scarcity of food-

and feed-grains, the opportunities for the small breeding organization to make a significant contribution appears enormous.

Botstein et al. (1980) proposed the development of restriction fragment length polymorphism-based linkage maps in *Homo sapiens*. Although his data were limited, sufficient preliminary information was derived by 1987 (Blake & Kleinhofs, 1988) to rationally propose the same venture in barley. The North American Barley Genome Mapping Project (NABGMP) grew out of a broadly-based collaborative effort to create a practically worthwhile linkage map in barley, and to identify the location of agronomically significant genes that exhibit allelic variation. The first population studied involved a cross between the varieties 'Morex' (a malting barley line developed by Don Rasmussen) and 'Steptoe', a variety developed by Bob Nilan with excellent Western adaptation but notoriously poor grain quality. Pat Hayes (Oregon State University) derived the 150 doubled haploid lines used for map construction and quantified trait loci (QTL) analysis, and now >40 locations have provided data on the performance of these lines (in Graingenes: http://wheat.usda.pw.gov under several headings). The Montana barley genetics program remains a small, but significant provider of information to the NABGMP. In exchange for the relatively small service of replicated field trials and a small portion of the markers on the map, we receive access to the information generated by the NABGMP.

BLAKE'S FIRST RULE

If somebody else wants to do your work for you, let them and thank them.

I gratefully acknowledge the contributions of my colleagues who participate in the NABGMP. From the Montana perspective, the most crucial information produced by the NABGMP was the elucidation of the genetics underlying the poor adaptation of Morex to Montana environments (Kleinhofs et al., 1993; Hayes et al., 1994; Larson et al., 1996). Hopeful crosses between Steptoe (and other Coast-class barleys) and Morex (and more recent products of the Minnesota barley improvement program) have been made by myself and others since the early 1970s to little effect. While recovering yield potential is not terribly difficult, recovering yield potential and malting quality in Western environments has not yet been achieved.

The initial results of the Steptoe–Morex QTL analysis demonstrated that one or two genes reside on chromosome 3 that have a profound impact on local adaptation (Fig. 8–1). As these QTL scans demonstrate, one major (and perhaps one lesser) gene on chromosome 3 have an impact on yield in arid western environments, while having no discernable impact on productivity in Minnesota, North Dakota, or Corvallis, Oregon. Although many characters were evaluated at many locations, the chromosome 3 yield effect showed neither a negative pleiotropic interaction nor a negative linkage effect with genes modifying any other measured character. This appears to be a relatively *safe* gene to substitute. Molecular marker analysis, when coupled with relatively low cost yield trial evaluation, can help identify genes associated with location-specific (G × E) interac-

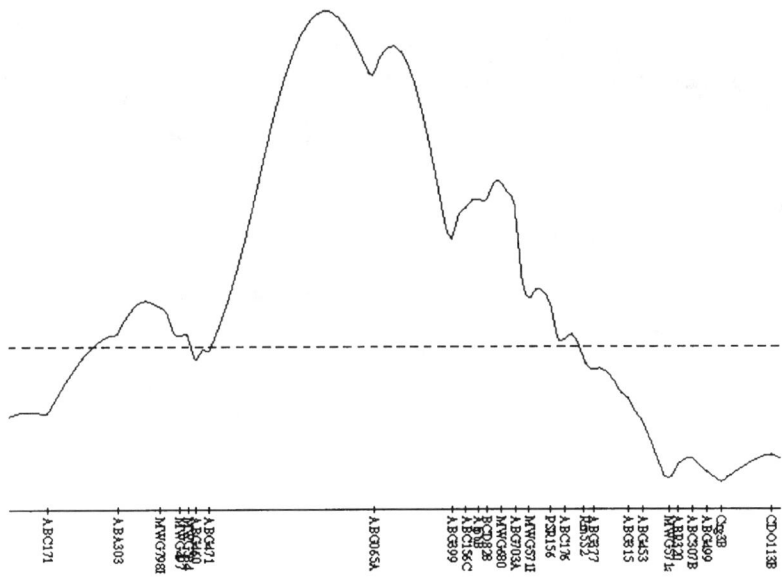

Fig. 8–1. Barley chromosome 3 from the Steptoe–Morex population. The upper quantitative trait loci (QTL) tracing shows the mean response over seven location years in Montana. The lower trace shows the response from one Oregon location. Genotype × environment interaction is key to the expression of this yield gene.

tions. Larson et al. (1996, 1997) demonstrated the value of this single gene substitution.

It had become apparent early in development of the NABGMP maps that RFLP analysis was too slow and skill-demanding to be a useful breeding tool in my small plant improvement program. A graduate student accepted the task of developing PCR derivatives of appropriate RFLP markers that might prove informative and straightforward to use in a backcrossing context. Steve Larson developed the markers, and began the process of moving the *Ras* (rachis stability) gene from the Steptoe background into the Morex background. Tables 8–1 and 8–2 show the result of this activity. Across five location × years replicated trials, backcross lines carrying the Steptoe allele outperformed those carrying the Morex allele at *Ras* by over 12% (Larson et al., 1996). The Great Western Malting Company performed malt analyses on these lines and found no significant interactions between the *RasA* gene and any of the characteristics significant to malting or brewing (Larson et al., 1997).

Our conclusion from these experiments is that capturing a large portion of the agronomic value of Steptoe in a Morex background may be accomplished relatively simply. Whether Montana can become a reasonable alternative to Minnesota, North Dakota, and Saskatchewan in the 6-rowed malting barley market remains to be seen; however, from the breeder–geneticist's perspective, the potential is enticing.

Table 8–1. 1987–1996 Spring barley multiyear summary report.

Identification	Pedigree	Locations	Yield
			kg ha^{-1}
PI491534	Gallatin (Check)	101	4648
CI 15514	Hector	98	4320
CI 15856	Lewis	101	4622
PI591823	Chinook	42	4685
SK 76333	Harrington	101	4489
ND 9866	Stark	71	4807
PI568246	Baronesse	76	5183
CI 15229	Steptoe	101	4918
CI 15773	Morex	75	3896

Table 8–2. Yield results from 1996 Northern Agriculture Research Center Intrastate Barley Yield Trial.

Variety	Yield
	kg ha^{-1}
CI 15229 Steptoe	3254
PI 491534 Gallatin	3160
ND 9866 Stark	3159
PI 591823 Chinook	3141
CI 15856 Lewis	3035
CI 15514 Hector	2758
SK 76333 Harrington	2781
CI 15773 Morex	2904
PI 568246 Baronesse	2756

SOLVING BARONESSE PROBLEM

In 1991 a we recommended that Montana barley producers grow a new 2-rowed feed barley developed by 'Nordstadt', a German plant breeding company. A local plant breeding company, Western Plant Breeders, arranged to market this variety. It continues to be a remarkable success. This variety is every breeders' dream, superb yield potential and broad adaptation. Grain quality is a problem, especially with respect to malting, but its feedlot performance is adequate and testweight is generally high. As a consequence, this variety has spread across the Pacific Northwest, and it has the potential to become the predominant regional variety.

In addition to being essentially unmaltable, this variety has defects associated with drought tolerance. Having been introduced in the relatively wet years of 1992 to 1995, it remained largely unexposed to traditional dryland stress. The past cropping season of 1996, while an above average production year statewide, was erratic in terms of statewide moisture distribution. Several locations demonstrated Baronesse's limitations relative to yield stability.

TECHNICAL PROBLEM AND ITS SOLUTION

Baronesse derives from essentially the same germplasm pool (the Northern European Hannchen group) as do all of our Pacific Northwest 2-rowed varieties.

While one-half of our STS markers (Blake et al., 1996) show polymorphism in the Steptoe–Morex population, only 10 produced reliable polymorphisms in a cross between Lewis, a 2-rowed well-adapted Montana product and Baronesse. We developed an RIL population in the hope that technical innovations would provide informative, useful markers.

Performance data were gathered from this RIL population in 1995 with and without supplemental irrigation at Bozeman, MT, and in 1996 with supplemental irrigation at Bozeman and Sidney, MT, and without supplemental irrigation at Huntley and Bozeman, MT. Data from these experiments are available at our website (http://hordeum.oscs.montana.edu).

AMPLIFIES FRAGMENT LENGTH POLYMORPHISM

Amplified Fragment Length Polymorphism (AFLP) analysis was developed and patented by Keygene (Vos et al., 1995). This technique uses an ingenious approach to scan the entire genome for restriction fragment length polymorphisms. While only about 10% of our STSs showed polymorphisms between Lewis and Baronesse, we were able to read on average nine AFLP polymorphisms per gel. In 2 wk we built a 116 point linkage map that we anchored to six of the seven barley chromosomes using our STS markers. Although achieving our initial objective of identifying the location genes impacting yield, when examined closely the map appeared to miss significant regions of the genome. Although seven linkage groups were obtained, four were near 100cM, one-half the size of the linkage groups described by Kleinhofs et al.(1993). A colleague (Charles Stuber) suggested that this might be the consequence of the pervasive problem of clustering of AFLP markers.

Messeguer et al. (1991) found that in tomato [*Lycopersicon lycopersicum* (L.) Karsten] unmethylated CG sites tended to be concentrated in high recombination frequency chromosomal regions, while their ^{m}CG counterparts tented to be more frequently found in recombinationally suppressed regions of the genome. Timmermans et al. (1996) further observed that recombination junctions cloned from maize (*Zea mays* L.) were free of cytosine methylation. More recently, Powell et al. (1997) reported that methylation sensitive *Pst1*, when used with *Mse1* in AFLP analysis provided markers in the recombinationally active, distal regions of barley chromosomes, while the methylation insensitive *EcoR1/Mse1* restriction endonuclease combination provided a high frequency of markers clustered near the barley centromere.

The *standard* AFLP kit (marketed by several providers, including Gibco and Perkin- Elmer) uses the restriction endonucleases *EcoR1* and *Mse1* to generate genomic restriction fragments that are then ligated to synthetic linkers. The linker sequences provide target sites for PCR amplification, and the complexity of the amplified products is limited to a useful level through the use of a number of selective bases, in a fashion analogous to that used in differential display. Neither *EcoR1* nor *Mse1* are particularly sensitive to target site methylation. *MspI* and *HpaII* provide an isoschizomer pair that differ in their ability to restrict methylated CCGG sequences. *MspII* restricts C^{m}CGG while *HpaII* does not. In a

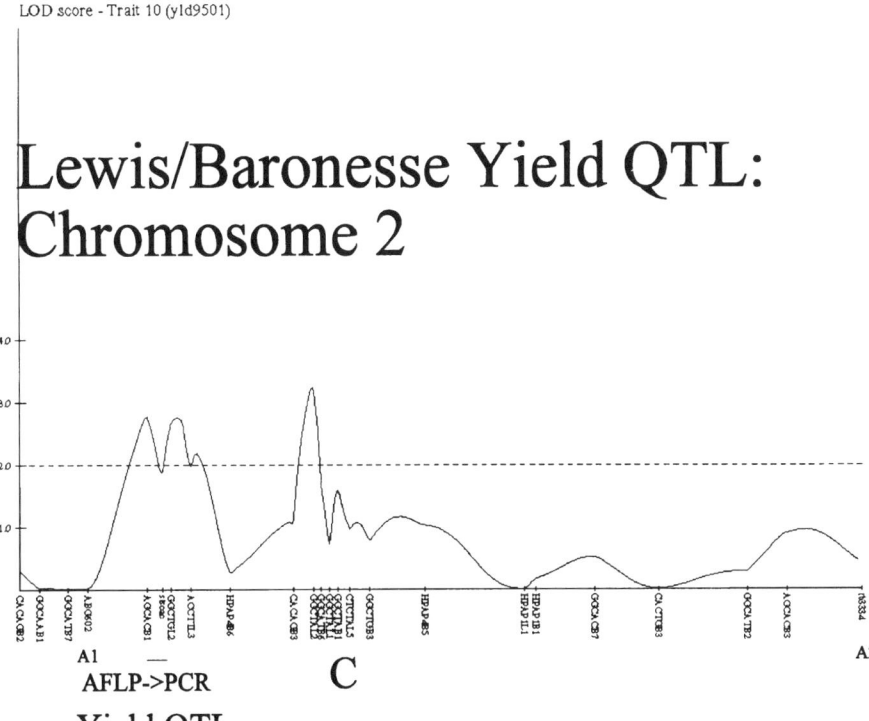

Fig. 8–2. Chromosome 2 yield quantitative trait loci (QTL). The chromosome is anchored at ABG602 and TB33/34, and the presumed centromeric cluster of unmethylated MspI restriction site polymorphisms is indicated by C. The converted sequence-tagged-site is labeled STSCAC, and the location of the AFLP polymorphism from which it derived is the adjacent ACCACB1. Markers whose names begin with *H* are derived from methylation-sensitive HpaII restriction site polymorphisms.

companion article (Kanazin et al., 1998, unpublished data) we demonstrate that C^mCGG sites are most commonly found in the recombinationally suppressed regions surrounding the centromeres of barley chromosomes, while unmethylated CCGG sequences tend to be found in recombinationally active chromosomal regions. This can be easily observed in Fig. 8–2. Addition of *HpaII* restriction site polymorphisms to the Lewis–Baronesse map expanded our linkage groups and improved the fit of STS anchors to their previously described positions.

RESULTS OF THE LEWIS–BARONESSE EXPERIMENT

This relatively simple field experiment, grown at six locations during 2 yr led to an improved understanding of the genetic differences delimiting our current working germplasm group and the remarkable introduction, Baronesse. If accurate, these results suggest that approximately 80% of the heritable variance for yield can be attributed to the action of genes or gene clusters on barley chro-

Table 8–3. The effect of quanatative trait loci (QTL) linked to STSCACB1 on yield and plant morphology in a population derived from a cross between 'Clark' and 'Baronesse.'

	Peduncle length	Plant height	Grain yield
	cm		kg ha^{-1}
Mean of lines with 'Clark' allele	20.1	84	7148
Mean of lines with 'Baronesse' allele	19.2	81	7490
P value	0.3	0.03	0.07

mosomes 2, 3 and 6. Backcrossing is now underway to move these genes into genetic backgrounds with improved malt potential, and enhanced feedlot performance.

AFLP analysis per se did not in our hands seem to be an ideal technology for single gene introgression. Sequencing gels and radioisotopic detection make this technique time-consuming and expensive; however, we found that cloning specific AFLP products was relatively simple (Kanazin et al., 1998, unpublished data), and conversion isolated, cloned AFLP bands to their single locus PCR derivatives was relatively straightforward. One of the AFLP markers in the chromosome 2 yield QTL region was cloned, sequenced, and converted to its single locus PCR counterpart. This primer pair was then used in segregation analysis, and given the name STSCACB1. One apparent recombinant was observed in the Lewis–Baronesse population between the AFLP marker from which STSCACB1 derived and the sequence-tagged-site derivative. I attribute that difference to error, although other explanations also are reasonable. To determine whether this marker would be productive in crosses between Pacific Northwest germplasm and Baronesse, we tested 60 lines derived from a cross between Baronesse and Clark (CI15657), which also were grown in a two replication randomized complete block design at the A.H. Post Research Farm, near Bozeman, MT, in 1996. The polymorphism detected in the Lewis–Baronesse population also was observed to segregate in this population, and an analysis of yield × STSCACB1 phenotype was performed. Lines carrying the Baronesse allele showed on average a 350 kg ha^{-1} yield advantage over those carrying the 'Clark' allele (Table 8–3). This result was very similar to that observed in the more thoroughly mapped Lewis–Baronesse population. This suggests that AFLP analysis, when done well, can provide adequate maps even in relatively narrow germplasm pools, and that when a marker is found to be linked to an important QTL, marker conversion is a reasonable objective.

MARKER TECHNOLOGY, HETEROSIS, AND SMALL RESEARCH PROJECTS

Many of the fundamental concepts of genetics are meaningless outside of context. Heterosis is only a meaningful term when use to describe the genetics underlying trait expression within a particular germplasm. Our initial experiments have failed to show that there exists an optimum barley genotype, but rather that several very productive barley genotypes exist, each of which performs differently in different environments. Optimizing economic and agronomic performance within environments seems more practically useful than attempting to identify the ultimate genotype.

Heterosis, per se, is either the result of dominant genes linked in repulsion or overdominance. While we in the inbred small grains have little opportunity to observe true overdominance, repulsion linkages are commonly observed. In the case of the Lewis–Baronesse population, an obvious repulsion linkage between a gene modifying head emergence and a gene affecting yield had an important impact on my breeding program. Early generation selection for head emergence is easy and reliable. In selecting for this highly heritable, observable character, we generally selected against a desirable yield factor. Now that markers are available that permit characterization of lines that carry genes for optimal yield, we can identify recombinants with nearly ideal plant height and peduncle extension.

When we began the quest for markers appropriate for marker-assisted-selection, our first need was informative markers. Several technologies proved useful, although RFLPs remained the dominant map-making tools throughout the early 1990s. Single locus PCR technologies have proven useful, although any technique that permits only one (or a handful) of loci to be scored per day is both tedious and expensive. AFLP analysis provides a multilocus approach to map construction while still using restriction site polymorphisms as its source of variation. Through careful selection of the types of restriction sites surveyed, it is likely that recombinationally normalized maps may be developed in many species. It appears as though barley is among these fortunate organisms.

As relatively inexpensive technologies like AFLP analysis become popularized across institutions and crops, questions regarding genetic mechanisms within specific contexts will be more commonly addressed, and more frequently answered. The next decade should be a lot of fun.

REFERENCES

Blake, T.K., D. Kadyrzhanova, K.W. Shepherd, A.K.M.R. Islam, P.L. Langridge, C.L. McDonald, J. Erpelding, S. Larson, N.K. Blake, and L.E. Talbert. 1996. STS-PCR markers appropriate for wheat–barley introgression. Theor. Appl. Genet. 93:826–832

Blake, T.K., and A. Kleinhofs. 1988. Applications of molecular biology to crop improvement. p. 991–999. *In* R.J. Summerfield (ed.) World crops: Cool season food legumes. Martinus Nijhoff, Dordrecht, the Netherlands.

Botstein, D, R.L. White, M. Skolnick, and R.W. Davis. 1980. Construction of a genetic linkage map in man using restriction fragment length polymorphisms. Am. J. Human Genet. 32:314–331.

Hayes, P.M., H. Cjirono, M. Witsenboer, M. Kuiper, M. Zabeau, K. Sato, A. Kleinhofs, D. Kudrna, A. Kilian, M. Saghai-Maroof, and D. Hoffman. 1997. Characterizing and exploiting genetic diversity and quantitative traits in barley (*Hordeum vulgare*) using AFLP markers. J. Quantitative Trait Loci. Vol. 3. http://probe.nalusda/gov:800/otherdocs/jqtl.

Hayes, P.M., B.H. Liu, S.J. Knapp, F. Chen, B. Jones, T. Blake, J. Frankowiak, D. Rasmussen, M. Sorrells, S.E. Ullrich, D. Wesenberg, and A. Kleinhofs. 1994. Quantitative trait locus effects and environmental interaction in a sample of North American barley germplasm. Theor. Appl. Genet. 87:397–401.

Kleinhofs, A., A. Kilian, M.A. Saghai-Maroof, R.M. Biyashev, P. Hayes, F.Q. Chen, N. Lapitan, A. Fenwick, T.K. Blake, V. Kanazin, E. Ananiev, L. Dahleen, C. Kudrna, J. Bollinger, S.J. Knapp, M. Sorrells, M. Heun, J.D. Franckowiak, D. Hoffman, R. Skadsen, and B.J. Steffenson. 1993. A molecular, isozyme and morphological map of the barley (*Hordeum vulgare*) genome. Theor. Appl. Genet. 86:705–712.

Larson, S., D. Habernicht, T. Blake, and M. Adamson. 1997. Backcross gains for six-rowed grain and malt qualities sith introgression of a feed barley yield QTL. J. Am. Soc. Brew. Chem. 55(2):52–57.

Larson, S., C. McDonald, T.K. Blake. 1996. Evaluation of barley chromosome 3 yield QTL in a backcross F_2 Population using PCR-STS markers. Theor. Appl. Genet. 93:618–625.

Martin, J.M., T.K. Blake, and E.A. Hockett. 1991. Diversity among North American spring barley cultivars based upon coefficients of parentage. Crop Sci. 31:1131–1137.

Messeguer, R., M.W. Ganal, J.C. Steffens, and S.D. Tanksley. 1991. Characterization of the level, target sites and inheritance of cytosine methylation in tomato nuclear DNA. Plant Mol. Biol. 16:753–770.

Powell, L., W.T.B. Thomas, E. Baird, P. Lawrence, A. Booth, B. Harrower, J.W. McNicol, and R. Waugh. 1997. Analysis of quantitative traits in barley by the use of Amplified Fragment Length Polymorphisms. Heredity 79:48–59.

Timmermans, M.C.P., O.P. Das, and J. Messing. 1996. Characterization of a Meiotic crossover in maize identified by restriction fragment length polymorphism-based method. Genetics 143:1771–1783.

Vos, P., R. Hogers, M. Bleeker, M. Reijans, T. van de Lee, M. Hornes, A. Ftijters, J. Pot, J. Peleman, M. Kuiper, and M. Zabeau. 1995. AFLP: A new technique for DNA fingerprinting. Nucl. Acids Res. 23:4407–4414.

9 Molecular Analyses and Heterosis in the Vegetables: Can We Breed Them Like Maize?

Michael J. Havey

USDA-ARS and Department of Horticulture
University of Wisconsin
Madison, Wisconsin

ABSTRACT

Advantages of hybrid-vegetable cultivars include superior performance over inbreds or open-pollinated populations, greater uniformity for maturity or quality, and the commercial benefit of annual seed sales. Hybrid cultivars have been developed for many vegetable species with different reproductive systems and are produced using large-scale emasculation, cytoplasmic-genic male sterility, and genetic or chemical modification of sex expression. In maize (*Zea mays* L.), hybrids generated by crossing among inbreds from different heterotic groups usually perform significantly better than those generated by crossing among inbreds within the same group. Genetic distance estimates based on molecular markers have been used to assign maize inbreds to previously characterized heterotic groups and occasionally correlated with single-cross performance. For crosses among more closely related inbreds, genetic distance estimates did not predict single-cross performance for grain or forage yields. The failure of molecular-based genetic distance estimates to predict superior hybrid performance also has been reported for oat (*Avena sativa* L.) and soybean [*Glycine max* (L.) Merr.]. Compared with maize, many vegetable crops have a restricted genetic background and molecular markers reveal relatively few polymorphisms. For cucumber (*Cucumis sativus* L.) and onion (*Allium cepa* L.), phenotypically distinct populations show a significant reduction in variability at molecular-marker loci. Vegetable crops with a narrow genetic background, such as cucumber, show little inbreeding depression and no significant heterosis. Although no direct comparisons between genetic-distance estimates and hybrid performance in vegetables have been reported, I expect that the narrow genetic background of many vegetable crops will restrict or preclude the identification of naturally occurring heterotic groups.

DEVELOPMENT OF HYBRID VEGETABLES

Soon after the initial descriptions in maize of the deleterious effects of inbreeding and recovery of vigor after crossing among inbreds, researchers applied these procedures to vegetable crops. Jones (1916) and Hayes and Jones

---

Copyright © 1998 Crop Science Society of America, 677 S. Segoe Rd., Madison, WI 53711, USA. *Concepts and Breeding of Heterosis in Crop Plants.* CSSA Special Publication no. 25.

(1916) described the performance of interpopulation crosses among locally grown tomato [*Lycopersicon lycopersicum* (L.) Karsten] and cucumber strains. They noted little to no heterosis when the parental populations were phenotypically similar. Today hybrid cultivars are available and represent a significant portion of commercial production for many naturally outcrossing vegetables, including sweet corn, carrot (*Daucus carota* L.), cucumber, onion, *Brassica oleracea* L. (cauliflower, broccoli, and cabbage; here-after referred to as the vegetable Brassicas), and asparagus (*Asparagus officinalis* L.), and the primarily self-pollinated tomato and pepper (*Capsicum annuum* L.; see Janick, 1998, this publication). The predominance of hybrid vegetables can be attributed to significant heterosis over the inbred parents or the original open pollinated population, greater uniformity at harvest, the ability to combine dominantly inherited disease resistances in the hybrid, high quality disease-free hybrid seed (as opposed to locally grown open pollinated seed), or the perception of superior hybrid performance. As agriculture became more specialized, growers eliminated poor quality or disease-infested seed as a constraint to production. Although I have no data on the relative importance of perceived superior hybrid performance or greater uniformity during the development of the hybrid vegetable seed industry, I believe these perceptions were a significant component, in addition to documented hybrid vigor or better seed quality.

My goal in this chapter was to review pertinent literature on relationships between genetic diversity among or within vegetable populations and hybrid performance. My comments will be limited to naturally outcrossing vegetables because I wish to make direct comparisons using maize (not sweet corn) as the standard. Although significant heterosis has been observed when crossing among naturally occurring populations or consciously selected inbreds of self pollinated vegetables (e.g., tomato or pepper), their performance would be better compared with the heterosis observed when crossing among homozygous lines of barley (*Hordeum vulgare* L.), bean (*Phaseolus vulgaris* L.), oat, pea (*Pisum sativum* L.), soybean, or wheat (*Triticum aestivum* L.).

INBREEDING DEPRESSION AND HETEROSIS

An array of systems enforce or encourage outcrossing in populations of the cross pollinated vegetables, including dioecy (asparagus), monoecy (cucumber and maize), sporophytic self incompatibility (the vegetable Brassicas), protandry (carrot and onion), and cytoplasmic-genic male sterility (carrot, maize, and onion). These same systems have been exploited successfully to commercially produce hybrid vegetable seed. Cytoplasmic-genic male sterility is or has been widely used to produce hybrid carrot, maize, onion, and the vegetable Brassicas; genetic dioecy in asparagus and cucumber; and sporophytic self incompatibility in the Brassicas.

Like maize, a significant reduction in vigor is observed when open pollinated populations of most outcrossing vegetable crops are subjected to inbreeding. Direct comparisons of inbreeding depression among different vegetables are difficult because of the plethora of yield and quality traits measured by different researchers (e.g., weight per area, individual fruit or plant size or weight, number of fruit per plant, earliness, uniformity, or storage ability). Percentage of reduction in vigor per generation over at least five generations of self pollination from open pollinated populations have been reported at 5% for asparagus (Ito & Currence, 1965), 9% for onion (Jones & Davis, 1944), and 0% for cucumber (Rubino & Wehner, 1986). Except for cucumber, these results are comparable to the 7% reduction in vigor reported by Jones (1924) for open pollinated populations of maize. The first inbred lines developed by vegetable breeders rarely passed the S_5 generation because of severe inbreeding depression and the longer generation time of biennial vegetable crops.

Crossing among these first inbred lines produced hybrids showing midparent and often high parent heterosis (Hutchins, 1938; Jones & Davis, 1944; Legg & Souther, 1967; Ito & Currence, 1965; Borchers, 1968); however, the performance of these initial hybrids was most often compared with that of the inbreds and not with the performance of the open pollinated population from which the inbreds were extracted. For example, Jones and Davis (1944) described experimental hybrids among onion inbreds distinguishable by different bulb colors or shapes. For six hybrids (disregarding reciprocal crosses), four showed both significant midparent and high parent heterosis; however, no comparisons were made with the open pollinated parental populations per se.

CONCEPT OF HETEROTIC GROUPS

The concept of heterotic groups was first developed in maize based on the observation that inbreds selected out of certain populations tended to produce better performing hybrids when crossed to inbreds from other groups (Hallauer et al., 1988). This recognition resulted from the systematic crossing of thousands of inbred lines from different source populations and evaluation of the hybrids. Two heterotic groups from which inbreds commonly are selected and used to produce superior maize hybrids are Iowa Stiff Stalk Synthetic (BSSS) and derivatives of Lancaster SureCrop (LSC; Darrah & Zuber, 1986; Gerdes & Tracy, 1993). Although both populations are primarily comprised of southern dent germplasm, LSC has more northern flint germplasm than BSSS (Smith, 1986; Gerdes & Tracy, 1993). A possible explanation for heterotic groups is that populations of divergent genetic backgrounds have unique allelic diversity that may have arisen from founder effects, genetic drift, or the accumulation of unique allelic diversity by mutation or selection. Significantly greater heterosis could result from this genetic diversity either by specific interallelic interactions (overdominance) or repulsion-phase linkages among loci showing dominance (pseudo-overdominance).

Divergence at molecular marker loci has not been a consistent predictor of single-cross performance, but has been useful in assigning inbreds to known heterotic groups and agreed with pedigree information (Lee et al., 1989; Melchinger et al., 1991; Messmer et al., 1992, 1993). This success reveals a potentially interesting application of molecular markers in vegetable crops. Because heterotic groups have not been described in the vegetables, could we use molecular markers to identify divergent populations of specific vegetables from which inbreds will combine significantly better? This approach represents the reverse of the process that led to the identification of heterotic groups in maize (i.e., development and testing of inbred lines from many populations, identification of heterotic groups, and establishment that heterotic groups are divergent at molecular marker loci).

DIVERSITY FOR MOLECULAR MARKERS AMONG PHENOTYPICALLY DIVERSE POPULATIONS

Different phenotypes may or may not reflect divergent genetic backgrounds. Phenotypically different populations may possess the same genetic background and divergent phenotypes may be conditioned by allelic differences at relatively few loci. In contrast, phenotypically different populations may possess divergent genetic backgrounds. Prior to the arrival of Europeans to North America, indigenous populations were growing two phenotypically divergent races of maize, Northern Flint and Southern Dent. Variability for isozymes and restriction fragment length polymorphisms (RFLPs) have demonstrated that these two endosperm types reflect divergent genetic backgrounds. Using open pollinated populations, Smith (1986) and Doebley et al. (1986, 1988) observed that Northern Flint and Southern Dent germplasms are divergent, the former originating from southwestern USA and northwestern Mexico and the latter from eastern Mexican dents. Midwestern Cornbelt Dents were derived from a mixing of dent and flint germplasm.

Vegetable crops show a wide diversity of phenotypes, selected either for specific market niches, production areas, or processing characteristics. Distinct phenotypic diversity exists among the cultivated types of the vegetable Brassicas (broccoli, Brussels sprouts, cabbage, cauliflower, or kale). Cucumber has been selected in northern regions for larger and more frequent tubercles that allow gas exchange during pickling; middle-eastern or Mediterranean cucumber populations possess few tubercles, are smoothed skinned, and consumed fresh (Lower & Edwards, 1986). The wild progenitor of onion was probably a rhizomorphous nonbulbing perennial plant and selection by humans produced an apically dominant biennial with bulbing controlled by length of night and minimum temperatures (Havey, 1993). Long-day onion populations are grown primarily at latitudes >40° and require 8 to 10 h of night to form bulbs. Short-day onions will form bulbs under lengths of night from 10 to 12 h. Long- and short-day onion germplasms have been maintained separately because long-day populations grown under short days will not bulb and short-day populations grown under long days produce small bulbs that often will not flower. Overall, selection of vegetables for production under specific environments or specific processing or market niches has produced numerous phenotypically diverse types.

Building on the model of maize, one could hypothesize that phenotypically diverse vegetable populations possess significant variability for molecular markers. I stress that this is a measure of the genetic background of populations and may not reflect variability at specific loci or chromosome regions conditioning or affecting heterosis. Open-pollinated maize populations have been characterized for isozymes; populations and inbreds have been characterized for both isozymes and restriction fragment length polymorphisms (RFLPs). Among older open-pollinated populations, 56 polymorphisms were revealed with 18 isozyme systems (Smith, 1986). Using multivariate analyses to visually display differences among populations, Smith (1986) demonstrated that Cornbelt dent germplasm was distributed around the southern dent type; northern flint types were more distantly removed from the dent populations. As expected, maize inbreds show greater numbers of polymorphisms than that observed among highly heterozygous populations, e.g., an average of 3.3 and 4.5 polymorphic bands or fragments for polymorphic isozymes (Kahler et al., 1986) and RFLPs (Melchinger et al., 1990), respectively. Variability for molecular markers generally agreed with pedigree information and assignment (based on hybrid performance) to known heterotic groups (Smith et al., 1990; Dudley et al., 1991; Melchinger et al., 1991); however, variability at molecular marker loci was not effective in predicting specific hybrid performance from crosses among maize inbreds (Lee et al., 1989; Melchinger et al., 1990, 1992). Molecular-based genetic distance estimates also failed to predict superior hybrid performance in oat (Moser & Lee, 1994) and soybean (Gizlice et al., 1993).

There are relatively few studies documenting variability for molecular markers among populations of the cross-pollinated vegetables. Nienhuis et al. (1993) identified RFLPs among commercial populations, hybrids, and inbreds of broccoli, cauliflower, and cabbage and demonstrated that these three cultivated types were clearly separated. Although variability for molecular markers was less than that observed in maize, the vegetable Brassicas possess significant levels of polymorphisms. In onion and cucumber a different picture has arisen, i.e., development of types for specific production or market niches is correlated with a significant reduction in genetic diversity. Bark and Havey (1995) identified random RFLPs among populations of long- and short-day onions. More genetic variability was observed among the short-day types; long-day storage populations were closely related and likely represent a derived group selected out of short-day germplasm. For all polymorphic probe-enzyme combinations, long-day populations showed a loss of a polymorphic fragment that was present in short-day germplasm. A reduction in genetic variability is even more evident in cucumber. North American and Mediterranean cucumber populations have a restricted genetic background as compared with onion, maize, or the vegetable Brassicas. This restricted genetic background is even more evident among populations of the European parthenocarpic slicing cucumber; 6 of 11 populations were identical for 40 probe-enzyme combinations known to detect RFLPs within cultivated cucumber germplasm (Dijkhuizen et al., 1996).

ANALYSES OF COMBINING ABILITY IN ONION

Inbred development in the vegetable crops is costly due to biennial generation times (leek and onion), strict vernalization requirements (beet, carrot, and the vegetable Brassicas), and difficulties in measuring quality characteristics. The recognition of heterotic groups in maize resulted from intercrossing of thousands of experimental inbreds. In any single vegetable crop, the numbers of combinations that have been evaluated must be at least one or two orders of magnitude fewer. As a result, vegetable breeders would benefit from a thorough understanding of the relationship between diversity for molecular markers and the degree of heterosis observed.

Although there are no direct comparisons between genetic distances estimated using isozymes or molecular markers and heterosis for any vegetable crop, I will compare variability for RFLPs with previously published descriptions of general (GCA) and specific (SCA) combining abilities in onion. As I stated earlier, we demonstrated that populations of long-day storage onions are closely related (Bark & Havey, 1995). These populations are spring sown, fall harvested, and stored for marketing throughout the winter months. They possess a restricted genetic background and represent a derived state selected from short-day populations for production under longer days (Bark & Havey, 1995). Hosfield et al. (1976, 1977a,b) evaluated a series of diallel crossing experiments for yield components and bulb quality. All inbreds used to generate three diallels were selected from long-day storage populations. Hosfield et al. (1976, 1977a,b) reported significant GCA and SCA effects for all traits, but GCA effects was always larger than SCA effects.

We evaluated the performance of testcrosses between three long day storage F_1 male-sterile lines and open pollinated long day storage and Spanish populations (Havey & Randle, 1996). The long-day Spanish populations were grown in Idaho, Oregon, and Washington and probably originated from crosses among long-day storage and short-day populations. Based on nuclear RFLPs, these Spanish populations have a more divergent genetic background compared with long-day storage populations (Bark & Havey, 1995). Male (GCA) effects were significant for yield, soluble solids content, and the proportion of larger bulbs. Male by female interactions (SCA) were significant for soluble solids content and storage ability. Overall, the divergent Spanish populations tend to combine better for larger bulbs with lower soluble solids and poorer storage ability. Testcrosses using three storage populations yielded significantly better than the open pollinated population per se (Havey & Randle, 1996). Overall, these studies of combining ability in onion using inbreds (Hosfield et al., 1976, 1977a,b; Joshi & Tandon, 1976) and open pollinated populations (Havey & Randle, 1996) have revealed that GCA effects are much more important than SCA. These results are similar to maize, in which GCA is more important than SCA for crosses among inbreds from different heterotic groups (see Melchinger, 1998, this publication).

REVIEW AND CONCLUSIONS

This review attempts to draw direct comparisons between out-crossing vegetable crops and maize for the existence of heterotic groups and the ability of molecular markers to identify these groups. For these vegetables, heterosis over the performance of inbred parents is common, but reports of significantly better hybrid performances over that of the open pollinated populations are rare.

Divergence for molecular markers correlates with phenotypic variability in maize and the vegetable Brassicas. But divergent phenotypic types are associated with a loss of genetic diversity in cucumber and onion. In my opinion, the restricted genetic backgrounds of some vegetable crops, such as cucumber and onion, reduce the number of repulsion-phase linkages among dominant loci and diminish SCA. Vegetables with a narrow genetic background may benefit from crossing with more divergent types to increase the diversity among populations. Subjecting these populations to recurrent selection should produce superior-combining inbreds, like that observed for BSSS and BSCB1 (Labate et al., 1997). Although we may not identify heterotic groups in the out-crossing vegetables, we can use lessons from maize to understand better individual vegetable crops to develop more effectively superior hybrids.

REFERENCES

Bark, O.H., and M.J. Havey. 1995. Similarities and relationships among open-pollinated populations of the bulb onion as estimated by nuclear RFLPs. Theor. Appl. Genet. 90:607–614.

Borchers, E.A. 1968. Broccoli inbreds, hybrids, and varieties. Proc. Am. Soc. Hort. Sci. 93:352–355.

Darrah, L.L., and M.S. Zuber. 1986. 1985 United States maize germplasm base and commercial breeding strategies. Crop Sci. 26:1109–1113.

Dijkhuizen, A., W.C. Kennard, M.J. Havey, and J.E. Staub. 1996. RFLP variability and genetic relationships in cultivated cucumber. Euphytica 90:79–87.

Doebley, J.F., M.M. Goodman, and C.W. Stuber. 1986. Exceptional genetic divergence of northern flint corn. Am. J. Bot. 73:64–69.

Doebley, J.F., J.D. Wendel, J.S.C. Smith, C.W. Stuber, and M.M. Goodman. 1988. The origin of Cornbelt maize: The isozyme evidence. Econ. Bot. 42:120–131.

Dudley, J.W., M.A. Saghai-Maroof, and G.K. Rufener. 1991. Molecular markers and grouping of parents in maize breeding programs. Crop Sci. 31:718–723.

Gerdes, J.T., and W.F. Tracy. 1993. Pedigree diversity within the Lancaster Surecrop heterotic group of maize. Crop Sci. 33:334–337.

Gizlice, Z., T.E. Carter, Jr., and J.W. Burton. 1993. Genetic diversity in North American soybean: II. Prediction of heterosis in F_2 populations of southern founding stock using genetic similarity measures. Crop Sci. 33:620–626.

Hallauer, A.R., W.A. Russell, and K.R. Lamkey. 1988. Corn breeding. p. 463–564. In G.F. Sprague and J.W. Dudley (ed.) Corn and corn improvement. 3rd ed. Agron. Monogr. 18. ASA, CSSA, and SSSA, Madison, WI.

Havey, M.J. 1993. Onion breeding. p. 35–49. In G. Kalloo and B. Berg (ed.) Genetic improvement of vegetable crops. Pergamon, Oxford.

Havey, M.J., and W.T. Randle. 1996. Combining abilities for yield and bulb quality among long- and intermediate-day open-pollinated onion populations. J. Am. Soc. Hort. Sci. 121:604–608.

Hayes, W., and D.F. Jones. 1916. First generation crosses in cucumbers. Rep. Connecticut Agric. Exp. Stn. 40:319–322.

Hosfield, G., G. Vest, and C. Peterson. 1976. A ten-parent diallel cross to evaluate inbred line performance and combining ability in onions. J. Am. Soc. Hort. Sci. 101:324–329.

Hosfield, G., G. Vest, and C. Peterson. 1977a. A seven-parent diallel cross in onions to evaluate general and specific combining ability and their interaction with years and locations. J. Am. Soc. Hort. Sci. 102:56–61.

Hosfield, G., G. Vest, and C. Peterson. 1977b. Heterosis and combining ability in a diallel cross of onions. J. Am. Soc. Hort. Sci. 102:355–360.

Hutchins, A.E. 1938. Some examples of heterosis in the cucumber, *Cucumis sativus* L. Proc. Am. Soc. Hort. Sci. 36:660–664.

Ito, P.J., and T.M. Currence. 1965. Inbreeding and heterosis in asparagus. Proc. Am. Soc. Hort. Sci. 86:660–664.

Janick, J. 1998. Hybrids in horticultural crops. p. 45–56. In K.R. Lamkey and J.E. Staub (ed.) Concepts and breeding of heterosis in crop plants. CSSA Spec. Publ. 25. CSSA, Madison, WI.

Jones, D.F. 1916. The effects of cross and self fertilization in tomatoes. Rep. Connecticut Agric. Exp. Stn. 40:305–318.

Jones, D.F. 1924. The attainment of homozygosity in inbred strains of maize. Genetics 9:405–418.

Jones, H., and G. Davis. 1944. Inbreeding and heterosis and their relation to the development of new varieties of onions. USDA Tech. Bull. No. 874. USDA, Washington, DC.

Joshi, H., and J. Tandon. 1976. Heterosis for yield and its genetic basis in onion. Indian J. Agric. Sci. 46:88–92.

Kahler, A.L., A.R. Hallauer, and C.O. Gardner. 1986. Allozyme polymorphisms within and among open-pollinated and adapted exotic populations of maize. Theor. Appl. Genet. 72:592–601.

Labate, J.A., K.R. Lamkey, M. Lee, and W.L. Woodman. 1997. Molecular genetic diversity after reciprocal recurrent selection in BSSS and BSCB1 maize populations. Crop Sci. 37:416–423.

Lee, M., E.B. Godshalk, K.R. Lamkey, and W.W. Woodman. 1989. Association of restriction fragment length polymorphisms among maize inbreds with agronomic performance of their crosses. Crop Sci. 29:1067–1071.

Legg, P.D., and F.D. Souther. 1967. Heterosis in intervarietal crosses in broccoli (*Brassica oleracea* var. *italica*). Proc. Am. Soc. Hort. Sci. 92:432–437.

Lower, R.L., and M.D. Edwards. 1986. Cucumber breeding. p. 173–207. In M.J. Bassett (ed.) Breeding vegetable crops. AVI Press, Westport, CT.

Melchinger, A.E., J. Boppenmaier, B.S. Dhillon, W.G. Pollmer, and R.G. Herrmann. 1992. Genetic diversity for RFLPs in European maize inbreds: II. Relation to performance of hybrids within versus between heterotic groups for forage traits. Theor. Appl. Genet. 84:627–681.

Melchinger, A.E., and R.K. Gumber. 1998. Overview of heterosis and heterotic groups in agronomic crops. p. 29–44. In K.R. Lamkey and J.E. Staub (ed.) Concepts and breeding of heterosis in crop plants. CSSA Spec. Publ. 25. CSSA, Madison, WI.

Melchinger, A.E., M. Lee, K.R. Lamkey, and W.L. Woodman. 1990. Genetic diversity for restriction fragment length polymorphisms: relation to estimated genetic effects in maize inbreds. Crop Sci. 30:1033–1040.

Melchinger, A.E., M.M. Messmer, M. Lee, W.L. Woodman, and K.R. Lamkey. 1991. Diversity and relationships among U.S. maize inbreds revealed by restriction fragment length polymorphism. Crop Sci. 31:669–678.

Messmer, M.M., A.E. Melchinger, J. Boppenmaier, E. Brunklaus-Jung, and R.G. Herrmann. 1992. Relationships among early European maize inbreds: I. Genetic diversity among flint and dent lines revealed by RFLPs. Crop Sci. 32:1301–1309.

Messmer, M.M., A.E. Melchinger, R.G. Herrmann, and J. Boppenmaier. 1993. Relationships among early European maize inbreds: II. Comparison of pedigree and RFLP data. Crop Sci. 33:944–950.

Moser, H., and M. Lee. 1994. RFLP variation and genealogical distance, multivariate distance, heterosis, and genetic variance in oats. Theor. Appl. Genet. 87:947–956.

Nienhuis, J., M.K. Slocum, D.A. DeVos, and R. Muren. 1993. Genetic similarity among *Brassica oleracea* L. genotypes as measured by restriction fragment length polymorphisms. J. Am. Soc. Hort. Sci. 118:298–303.

Rubino, D.B., and T.C. Wehner. 1986. Effect of inbreeding on horticultural performance of lines developed from an open pollinated pickling cucumber population. Euphytica 35:459–464.

Smith, J.S.C. 1986. Genetic diversity within the corn belt dent racial complex of maize (*Zea mays* L.). Maydica 21:349–367.

Smith, O.S., J.S.C. Smith, S.L. Bowen, R.A. Tenborg, and S.J. Wall. 1990. Similarities among a group of elite maize inbreds as measured by pedigree, F1 grain yield, grain yield heterosis, and RFLPs. Theor. Appl. Genet. 80:833–840.

10 Predicting the Performance of Untested Single Crosses: Trait and Marker Data

Rex Bernardo

Limagrain Genetics
Champaign, Illinois

ABSTRACT

Predicting the performance of untested single crosses is an important objective in hybrid breeding programs. The effectiveness of best linear unbiased prediction (BLUP) of grain yield, moisture, stalk lodging, and root lodging of untested maize (*Zea mays* L.) single crosses was investigated. Multilocation data, from 1990 to 1995, for 4775 single crosses were obtained from the hybrid testing program of Limagrain Genetics. For each of 16 heterotic patterns, the performance of m untested single crosses was predicted from the trait phenotypes (T-BLUP) of n tested single crosses as $\mathbf{y_M} = \mathbf{C_{MP}} \mathbf{C_{PP}}^{-1} \mathbf{y_P}$, where $\mathbf{y_M} = m \times 1$ vector of predicted performance of untested single crosses; $\mathbf{C_{MP}} = m \times n$ matrix of genetic covariances between untested single crosses and tested single crosses; $\mathbf{C_{PP}} = n \times n$ phenotypic covariance matrix among tested single crosses; and $\mathbf{y_P} = n \times 1$ vector of average performance of tested single crosses, corrected for yield trial effects. For one heterotic pattern with available parental RFLP data, performance also was predicted from (i) marker genotypes (M-BLUP) at loci with significant general combining ability effects and (ii) both trait phenotypes and marker genotypes (TM-BLUP). Correlations between predicted and observed performance were obtained by a delete-one cross-validation procedure. Across heterotic patterns, the correlations for T-BLUP ranged from 0.463 to 0.770 for yield, 0.868 to 0.936 for moisture, 0.466 to 0.685 for stalk lodging, and 0.164 to 0.518 for root lodging. Out of 74 RFLP loci, 9 had significant ($P \leq 0.1$) effects on yield, 14 on moisture, and 22 on stalk lodging. Correlations between predicted and observed yield were 0.764 for T-BLUP, 0.765 for TM-BLUP, and 0.341 for M-BLUP. Simplifying assumptions regarding marker effects and variances were used in TM-BLUP and M-BLUP, and whether a more complex model would lead to better TM-BLUP and M-BLUP predictions is yet to be determined. The results indicated that BLUP based on trait data alone is useful for routine identification of superior single crosses prior to field testing.

Commercial maize breeders routinely evaluate several hundred single crosses in their respective breeding programs each year. Typically, <1% of the tested single crosses eventually become commercial hybrids (Hallauer, 1990). Heterosis is the basis for the development of single-cross cultivars, but prediction of heterosis itself is not an important objective in commercial maize breeding programs.

Copyright © 1998 Crop Science Society of America, 677 S. Segoe Rd., Madison, WI 53711, USA.
Concepts and Breeding of Heterosis in Crop Plants. CSSA Special Publication no. 25.

Rather, breeders are interested in predicting single-cross performance prior to making the actual crosses and evaluating them in the field.

Best linear unbiased prediction based on trait phenotypes (T-BLUP) has been found useful for selection of untested maize single crosses. Correlations between predicted and observed performance, when the number of tested single crosses was >100, have ranged from 0.426 to 0.762 for yield, 0.754 to 0.933 for moisture, 0.300 to 0.739 for stalk lodging, and 0.164 to 0.532 for root lodging (Bernardo, 1996). An attractive feature of T-BLUP is that the predictions are made from data routinely generated in yield trials. Restriction fragment length polymorphism (RFLP) marker data on the parental inbreds of single crosses have become increasingly available in commercial maize breeding programs. If single cross trait data and parental marker data are available, BLUP may be useful for detecting markers with significant general combining ability (GCA) effects for different traits. Single-cross performance then may be predicted from (i) marker genotypes (M-BLUP) at loci with significant GCA effects or (ii) both trait phenotypes and marker genotypes (TM-BLUP). But the usefulness of T-BLUP, TM-BLUP, and M-BLUP has not been compared.

My objectives are to: (i) present results on the use of T-BLUP for predicting yield, moisture, stalk lodging, and root lodging with data sets typically encountered in a commercial maize hybrid breeding program; (ii) describe how BLUP may be used for routine detection of marker-trait associations; and (iii) compare the effectiveness of T-BLUP, TM-BLUP, and M-BLUP in predicting single-cross performance.

MATERIALS AND METHODS

Trait Data Sets

A total of 404 private inbreds were assigned to nine different heterotic groups, arbitrarily designated A, B, C, D, E, F, G, H, and I. Sixteen different cross combinations (i.e., heterotic patterns) between the nine heterotic groups were considered. Coefficients of coancestry between inbreds were obtained from pedigree records (but not from marker data) by tabular analysis (Emik & Terrill, 1949). The hybrid performance data sets comprised the results from multilocation yield trials, conducted by Limagrain Genetics from 1990 to 1995, of 4775 single crosses. The trait data sets were highly unbalanced across multilocation trials but, disregarding occasional missing plots at individual locations, balanced within multilocation trials. The performance at each individual location was not considered in the data analyses. Rather, the T-BLUP, TM-BLUP, and M-BLUP analyses were based upon the average performance of a single cross or check hybrid at several locations in a multilocation yield trial. Data were recorded for grain yield (t ha^{-1} at 155 g H$_2$O kg^{-1}), moisture (g kg^{-1}), and the percentage of stalk and root lodging.

Marker Data Sets

For 55 inbreds in heterotic Group B and 53 inbreds in heterotic Group D, patterns of hybridization fragments (bands) were determined for 74 well-dispersed probes and restriction digests of genomic DNA. Either *Eco*RI, *Eco*RV, or *Hind*III was used as the restriction enzyme in combination with each probe. Each of the 74 probes mapped to a single locus in previous screening and mapping experiments. Therefore, each of the 74 probe-enzyme combinations was considered an RFLP locus and each unique banding pattern an RFLP allele.

T-BLUP Data Analysis

Assume n single crosses were made between n_1 inbreds from heterotic group 1 and n_2 inbreds from heterotic group 2. The single crosses, along with n_C check hybrids, were evaluated in t different yield trials resulting in p total data points. The linear model in T-BLUP, assuming negligible epistasis, was:

$$\mathbf{y} = \mathbf{X}\boldsymbol{\beta} + \mathbf{Z}_0 \mathbf{c} + \mathbf{Z}_1 \mathbf{g}_1 + \mathbf{Z}_2 \mathbf{g}_2 + \mathbf{Z}\mathbf{d} + \mathbf{e} \quad [1]$$

where: $\mathbf{y} = p \times 1$ vector of observed performance for a given trait (i.e., hybrid by multilocation trial means); $\boldsymbol{\beta} = t \times 1$ vector of yield trial effects; $\mathbf{c} = n_C \times 1$ vector of check hybrid effects; $\mathbf{g}_1 = n_1 \times 1$ vector of GCA effects of inbreds in heterotic group 1; $\mathbf{g}_2 = n_2 \times 1$ vector of GCA effects of inbreds in heterotic group 2; $\mathbf{d} = n \times 1$ vector of specific combining ability (SCA) effects; $\mathbf{e} = p \times 1$ vector of residual effects; and $\mathbf{X}, \mathbf{Z}_0, \mathbf{Z}_1, \mathbf{Z}_2$, and $\mathbf{Z}$ were incidence matrices of 1s and 0s relating $\mathbf{y}$ to $\boldsymbol{\beta}, \mathbf{c}, \mathbf{g}_1, \mathbf{g}_2$, and $\mathbf{d}$, respectively. Multilocation yield trial effects were considered fixed whereas all other effects in the model were considered random. Because the data set comprised the average performance of single crosses and check hybrids across locations, the effects of the individual locations were not in the model, although locations were considered random.

Assume i and i' were two inbreds from heterotic group 1 whereas j and j' were two inbreds from heterotic group 2. Inbreds in heterotic group 1 were unrelated to inbreds in heterotic group 2. The covariance between $i \times j$ and $i' \times j'$ was (Stuber & Cockerham, 1966):

$$\text{Cov}[(i \times j),(i' \times j')] = f_{ii'} V_{GCA(1)} + f_{jj'} V_{GCA(2)} + f_{ii'} f_{jj'} V_{SCA} \quad [2]$$

where: $V_{GCA(1)}$ = summed (across loci) GCA variance of alleles from heterotic group 1; $V_{GCA(2)}$ = summed (across loci) GCA variance of alleles from heterotic group 2; V_{SCA} = summed (across loci) SCA variance of paired alleles from heterotic groups 1 and 2; $f_{ii'}$ = coefficient of coancestry between inbreds i and i'; and $f_{jj'}$ = coefficient of coancestry between inbreds j and j'. The random effect vectors $\mathbf{c}, \mathbf{g}_1, \mathbf{g}_2, \mathbf{d}$, and $\mathbf{e}$ had zero means and the following covariances:

$$\text{Var} \begin{bmatrix} \mathbf{c} \\ \mathbf{g}_1 \\ \mathbf{g}_2 \\ \mathbf{d} \\ \mathbf{e} \end{bmatrix} = \begin{bmatrix} \mathbf{I}V_C & 0 & 0 & 0 & 0 \\ 0 & \mathbf{G}_1 V_{GCA(1)} & 0 & 0 & 0 \\ 0 & 0 & \mathbf{G}_2 V_{GCA(2)} & 0 & 0 \\ 0 & 0 & 0 & \mathbf{D}V_{SCA} & 0 \\ 0 & 0 & 0 & 0 & \mathbf{R}V_R \end{bmatrix}$$

where V_C = variance among check hybrids and V_R = residual variance.

The matrix elements were equal to $f_{ii'}$ in $\mathbf{G_1}$, $f_{jj'}$ in $\mathbf{G_2}$, and $f_{ii'}f_{jj'}$ in $\mathbf{D}$. The check hybrids were assumed unrelated and the covariance of $\mathbf{c}$ was $\mathbf{I}V_C$, where $\mathbf{I}$ was an identity matrix. In $\mathbf{R}$, the off-diagonal elements were assumed zero and the ith diagonal element was the reciprocal of the number of observations (locations) for the ith data point in $\mathbf{y}$. For example, if the first element in $\mathbf{y}$ was the average performance of a single cross in a trial at four locations, the first diagonal element in $\mathbf{R}$ was 1/4.

Solutions to β, $\mathbf{c}$, $\mathbf{g_1}$, $\mathbf{g_2}$, and $\mathbf{d}$ were obtained by solving the mixed-model equations for predicting single-cross performance [see Bernardo, 1996, for details]. Estimates of residual and genetic variances were obtained iteratively by restricted maximum likelihood (Henderson, 1985). After estimates of yield trial effects (β) and residual and genetic variances were obtained, the performance of the $(n_1 n_2 - n) = m$ untested single crosses was predicted from the performance of the n tested single crosses. Let $\mathbf{y_P}$ be a $n \times 1$ vector of average performance of tested single crosses, corrected for yield trial effects (i.e., β):

$$\mathbf{y_P} = (\mathbf{Z'R^{-1}Z})^{-1} \mathbf{Z'R^{-1}}(\mathbf{y} - \mathbf{X}\beta).$$

The performance for each trait of the untested single crosses was predicted as:

$$\mathbf{y_M} = \mathbf{C_{MP}} \mathbf{C_{PP}}^{-1} \mathbf{y_P} \qquad [3]$$

where: $\mathbf{y_M} = m \times 1$ vector of predicted performance of the untested single crosses; $\mathbf{C_{MP}} = m \times n$ matrix of genetic covariances between the untested single crosses and the tested single crosses; and $\mathbf{C_{PP}} = n \times n$ phenotypic variance-covariance matrix among the tested single crosses. The elements of $\mathbf{C_{MP}}$ as well as the off-diagonal elements of $\mathbf{C_{PP}}$ were calculated using Eq. [2]. The ith diagonal element of $\mathbf{C_{PP}}$ was equal to the sum of $V_{GCA(1)}$, $V_{GCA(2)}$, V_{SCA}, and the quantity $V_R/(i$th diagonal element of the diagonal $\mathbf{Z'R^{-1}Z}$ matrix).

A delete-one cross-validation procedure (Efron, 1981) was performed to evaluate the usefulness of BLUP for predicting single-cross performance. Given n tested single crosses, the data for the first single cross were disregarded and predicted from the remaining $n - 1$ single crosses. The performance of the 2nd, 3rd, ..., nth single cross was likewise predicted from the remaining $n - 1$ single crosses. The correlation between the predicted and observed performance (sample size of n) was calculated for each trait as an indication of the effectiveness of T-BLUP. Values of predicted and observed performance were not necessarily obtained from independent sets of yield trials.

TM-BLUP and M-BLUP Data Analyses

The TM-BLUP and M-BLUP data analyses were performed only for the 464 single crosses in the B × D heterotic pattern. In TM-BLUP, the linear model in Eq. [1] was expanded to include marker GCA effects for a given trait:

$$\mathbf{y} = \mathbf{X}\beta + \mathbf{Z_0 c} + \mathbf{Z_1 g_1} + \Sigma_k \mathbf{Y_{1k} m_{1k}} + \mathbf{Z_2 g_2} + \Sigma_k \mathbf{Y_{2k} m_{2k}} + \mathbf{Z d} + \mathbf{e}$$

where: $\mathbf{m}_{1k} = a_k \times 1$ vector of random GCA effects of marker alleles at the kth RFLP locus ($k = 1$ to q) in heterotic group B; $\mathbf{m}_{2k} = a'_k \times 1$ vector of random GCA effects of marker alleles at the kth RFLP locus in heterotic group D; q = number of RFLP loci in the model; a_k = number of marker alleles at the kth RFLP locus in heterotic group B; a'_k = number of marker alleles at the kth RFLP locus in heterotic group D; and $\mathbf{Y}_{1k}$ and $\mathbf{Y}_{2k}$ were incidence matrices of 1s and 0s relating $\mathbf{y}$ to $\mathbf{m}_{1k}$ and $\mathbf{m}_{2k}$, respectively.

At a given RFLP locus, the numbers of marker alleles in each heterotic group were not necessarily equal. Also, marker GCA effects of the same allele at a given RFLP locus differed between the two heterotic groups (i.e., $\mathbf{m}_{1k} \neq \mathbf{m}_{2k}$). Normal distributions with a mean of zero and variances of $\text{Var}(\mathbf{m}_{1k}) = \mathbf{I}_k V_{\text{MGCA}(1)}$ and $\text{Var}(\mathbf{m}_{2k}) = \mathbf{I}_k V_{\text{MGCA}(2)}$ were assumed for marker GCA effects at the kth RFLP locus. A problem that arose was that a_k and a'_k were generally ≤ 5. Because variances estimated from a sample of five would not have been reliable, marker GCA variances were assumed homogeneous across RFLP loci in each heterotic group. Hence, for any of the q RFLP loci in the model, variances of marker GCA effects were $\text{Var}(\mathbf{m}_{1k}) = \mathbf{I} V_{\text{MGCA}(1)}$ and $\text{Var}(\mathbf{m}_{2k}) = \mathbf{I} V_{\text{MGCA}(2)}$. $V_{\text{MGCA}(1)}$ and $V_{\text{MGCA}(2)}$ were average marker GCA variances across RFLP loci in heterotic groups B and D, respectively.

The covariance between the $i \times j$ and $i' \times j'$ single crosses in TM-BLUP was obtained by expanding Eq. [2]:

$\text{Cov}[(i \times j),(i' \times j')]$

$$= f_{ii'} V_{\text{GCA}(1)} + V_{\text{MGCA}(1)} \Sigma_k h_{ii'} + f_{jj'} V_{\text{GCA}(2)}$$
$$+ V_{\text{MGCA}(2)} \Sigma_k h_{jj'} + f_{ii'} f_{jj'} V_{\text{SCA}} \qquad [4]$$

where $h_{ii'}$ (or $h_{jj'}$) was equal to 1 if the i and i' (or j and j') parental inbreds had the same marker allele at the kth RFLP locus. Otherwise, $h_{ii'}$ or $h_{jj'}$ was equal to zero. Values of $V_{\text{GCA}(1)}$ and $V_{\text{GCA}(2)}$ were not equal between the T-BLUP and TM-BLUP models.

Procedures for solving the mixed-model equations and estimating variances for TM-BLUP were similar to those used for T-BLUP. A procedure analogous to backwards elimination in multiple regression was used to detect RFLP loci with significant effects for each trait. All RFLP loci ($q = 74$) were initially used in TM-BLUP. Significance ($P \leq 0.1$) of each element in $\hat{\mathbf{m}}_{1k}$ and in $\hat{\mathbf{m}}_{2k}$ was tested by a z-test. The appropriate denominators in the z-tests corresponded to the diagonal elements of $\text{Var}(\hat{\mathbf{m}}_{1k}) = [\mathbf{I} V_{\text{MGCA}(1)} - V_R \mathbf{c}_{m1k,m1k}]$ and $\text{Var}(\hat{\mathbf{m}}_{2k}) = [\mathbf{I} V_{\text{MGCA}(2)} - V_R \mathbf{c}_{m2k,m2k}]$ (Henderson, 1985). The $\mathbf{c}_{m1k,m1k}$ and $\mathbf{c}_{m2k,m2k}$ matrices corresponded to $\mathbf{m}_{1k}$ and $\mathbf{m}_{2k}$, respectively, in the inverse of the coefficient matrix of the mixed-model equations. Only those RFLP loci with at least one allele having a significant effect, in either or both heterotic groups, were retained in subsequent analyses. The procedure was repeated until all the RFLP loci retained in TM-BLUP had significant marker GCA effects in either or both heterotic groups.

In the delete-one cross-validation procedure for TM-BLUP, the elements of $\mathbf{C}_{\text{MP}}$ and $\mathbf{C}_{\text{PP}}$ (see Eq. [3]) were calculated with Eq. [4]. The quantity $V_R/(i$th diagonal element of the diagonal $\mathbf{Z'R^{-1}Z}$ matrix) was added to the diagonal elements of $\mathbf{C_{PP}}$.

Only the RFLP loci with significant marker GCA effects (i.e., determined in TM-BLUP) were used for predicting single-cross performance in M-BLUP. Hence, the elements of $\mathbf{C_{MP}}$ and $\mathbf{C_{PP}}$ in M-BLUP were calculated as:

$$\text{Cov}[(i \times j),(i' \times j')] = V_{MGCA(1)} \Sigma_k h_{ii'} + V_{MGCA(2)} \Sigma_k h_{jj'}$$

Computer Software and Hardware

All the necessary computations for T-BLUP, TM-BLUP, and M-BLUP were done with *lgHYPER*, a proprietary FORTRAN 77 program. The *lgHYPER* program was compiled for 32-bit DOS with virtual memory support. All calculations were done on a 150-MHz Pentium (Intel Corporation) machine with 64 megabytes of RAM. The procedures were computer-intensive; for example, total computer run time for analysis of yield in the B × D heterotic pattern was about 12 h for T-BLUP and 107 h for TM-BLUP.

Table 10–1. T-BLUP variances and correlations between predicted and observed performance for yield (t ha^{-1}), moisture (g kg^{-1}), and stalk and root lodging (%) for 16 heterotic patterns (Group 1 × Group 2) in maize.

Group 1	Group 2	Number of single crosses		Yield	Moisture	Stalk lodging	Root lodging
A	D	288	V_{SCA}/V_G †	0.19	0.19	0.21	0.25
			Correlation ‡	0.678	0.890	0.578	0.516
			Upper bound §	0.784	0.951	0.791	0.729
A	F	232	V_{SCA}/V_G	0.50	0.09	0.34	0.60
			Correlation	0.642	0.936	0.662	0.440
			Upper bound	0.762	0.958	0.888	0.736
A	G	147	V_{SCA}/V_G	0.12	0.33	0.15	0.14
			Correlation	0.589	0.868	0.537	0.518
			Upper bound	0.819	0.957	0.834	0.801
A	H	206	V_{SCA}/V_G	0.46	0.13	0.29	0.45
			Correlation	0.536	0.889	0.603	0.419
			Upper bound	0.802	0.962	0.848	0.570
A	I	159	V_{SCA}/V_G	0.52	0.15	0.37	0.45
			Correlation	0.600	0.927	0.584	0.448
			Upper bound	0.785	0.962	0.827	0.595
B	D	464	V_{SCA}/V_G	0.08	0.06	0.14	0.75
			Correlation	0.764	0.934	0.641	0.498
			Upper bound	0.789	0.953	0.795	0.768
B	F	551	V_{SCA}/V_G	0.13	0.05	0.21	0.45
			Correlation	0.693	0.910	0.674	0.267
			Upper bound	0.745	0.957	0.808	0.657
B	G	236	V_{SCA}/V_G	0.25	0.16	0.19	0.29
			Correlation	0.463	0.876	0.572	0.290
			Upper bound	0.640	0.948	0.711	0.505

(continued on next page)

RESULTS AND DISCUSSION

T-BLUP Correlations between Predicted and Observed Performance

Across the 16 heterotic patterns, the correlation between the predicted and observed performance ranged from 0.463 to 0.770 for yield, 0.868 to 0.936 for moisture, 0.466 to 0.685 for stalk lodging, and 0.164 to 0.518 for root lodging (Table 10–1). As Bernardo (1996) reported, the square root of heritability for each trait sets a theoretical upper bound (Table 10–1) on the correlation between predicted and observed values and should be the reference for interpreting the correlations obtained with the cross-validation procedure.

For yield, most of the correlations between predicted and observed performance were ≥0.60 and were 75 to 85% of the corresponding theoretical upper bounds. These correlations were not perfect but seemed sufficiently high (Johnson, 1989; Bernardo, 1992) for pick-the-winner schemes that characterize hybrid testing programs. For example, suppose a breeder desires to identify the best single cross (i.e., with the highest genetic value) out of 100. A correlation

Table 10–1. Continued.

Group 1	Group 2	Number of single crosses		Yield	Moisture	Stalk lodging	Root lodging
B	H	486	V_{SCA}/V_G	0.09	0.05	0.28	0.03
			Correlation	0.641	0.901	0.555	0.242
			Upper bound	0.772	0.960	0.736	0.566
B	I	377	V_{SCA}/V_G	0.18	0.05	0.16	0.50
			Correlation	0.658	0.921	0.678	0.513
			Upper bound	0.787	0.964	0.840	0.501
C	D	255	V_{SCA}/V_G	0.11	0.05	0.35	0.11
			Correlation	0.770	0.929	0.535	0.380
			Upper bound	0.863	0.962	0.807	0.634
C	F	301	V_{SCA}/V_G	0.33	0.10	0.32	0.81
			Correlation	0.642	0.886	0.616	0.447
			Upper bound	0.778	0.955	0.887	0.807
C	G	202	V_{SCA}/V_G	0.18	0.13	0.34	0.49
			Correlation	0.655	0.896	0.506	0.164
			Upper bound	0.836	0.964	0.806	0.512
C	H	311	V_{SCA}/V_G	0.38	0.07	0.39	0.17
			Correlation	0.504	0.921	0.536	0.330
			Upper bound	0.744	0.966	0.799	0.539
C	I	208	V_{SCA}/V_G	0.29	0.09	0.05	0.39
			Correlation	0.661	0.879	0.685	0.413
			Upper bound	0.813	0.967	0.879	0.520
D	E	352	V_{SCA}/V_G	0.40	0.08	0.09	0.40
			Correlation	0.656	0.890	0.466	0.308
			Upper bound	0.810	0.940	0.667	0.632

† V_{SCA}/V_G = ratio of V_{SCA} to the total genetic variance ($V_G = V_{GCA(1)} + V_{GCA(2)} + V_{SCA}$).
‡ All correlation coefficients were significantly different from zero at $P \leq 0.05$.
§ Theoretical upper bound, equal to the square root of heritability, on the correlation between predicted and observed single-cross performance.

between predicted and true genetic value of 0.60 would allow a breeder to select the top 20 out of 100 single crosses while maintaining at least an 80% chance of retaining the best hybrid in the selected group (Bernardo, 1992).

Whereas the correlations between predicted and observed moisture were high (>0.85), such correlations varied in magnitude and were often low for root lodging (Table 10–1). From experience maize breeders know that field trial results for root lodging are much less consistent than those for moisture. For root lodging, most of the correlations ranged from ≈0.30 to 0.50 and were 50 to 90% of their corresponding upper bounds.

For yield, the average ratio across heterotic patterns of V_{SCA} to total genetic variance ($V_G = V_{GCA(1)} + V_{GCA(2)} + V_{SCA}$) was 0.26 (Table 10–1). In contrast, average V_{SCA}/V_G was 0.11 for moisture. In agreement with published results (Hallauer & Miranda, 1981), the importance of nonadditive effects relative to additive effects was less for moisture than for yield. Average V_{SCA}/V_G across heterotic patterns also was high for stalk lodging (0.24) and root lodging (0.39), indicating the importance of both additive and nonadditive effects for these traits.

Markers with Significant General Combining Ability Effects

Out of 74 RFLP loci, 9 had significant marker GCA effects on yield, 14 on moisture, and 22 on stalk lodging in the B × D heterotic pattern (Table 10–2). Only one RFLP locus with three alleles had significant effect on root lodging, and the results for this trait are not discussed further because of the resulting lack of precision in the estimates of $V_{MGCA(1)}$ and $V_{MGCA(2)}$. Some RFLP loci had significant effects in both the B and D heterotic groups. Other RFLP loci had marker alleles with significant GCA effects in one heterotic group only. For yield, heterotic group B had fewer RFLP loci with significant effects than heterotic group D. For moisture and stalk lodging, the number of RFLP loci with significant GCA effects was comparable between the two heterotic groups.

Names and map positions of the RFLP loci with significant GCA effects for each trait are not presented due to the proprietary nature of the data. Such information was not necessary for the use of markers to predict single-cross performance in TM-BLUP and M-BLUP. Each of the 10 chromosomes had RFLP loci with significant GCA effects for at least one trait. None of the chromosomes had a large cluster of RFLP loci affecting yield. In contrast, chromosomes 4 and 8 had three or more RFLP loci associated with moisture, whereas chromosomes 1, 4, and 5 had three of more RFLP loci associated with stalk lodging.

At many of the RFLP loci associated with each trait, marker alleles with a positive effect in one heterotic group had a corresponding negative (although nonsignificant) effect in the other heterotic group (results not shown). For example, one allele at an RFLP locus on chromosome 10 had a marker GCA effect of 0.039 t ha^{-1} in heterotic group B and -0.044 (significant at $P \leq 0.1$) t ha^{-1} in heterotic group D. Thus, generalization of GCA effects of marker alleles across heterotic groups may not be possible and, as with GCA effects of inbreds, a separate analysis of marker GCA effects would be necessary for each heterotic pattern.

The RFLP loci explained a substantial proportion of the GCA variance for each trait (Table 10–3). For example, for yield among inbreds in heterotic group

Table 10–2. Numbers of RFLP loci with significant ($P \leq 0.1$) general combining ability effects in the B × D maize heterotic pattern.

Trait	RFLP loci with significant effects	Significant effects in:		
		Both B and D	B only	D only
Yield	9	4	1	4
Moisture	14	2	5	7
Stalk lodging	22	6	8	8

B, each of the nine RFLP loci with significant effects had a variance of $V_{MGCA(1)} = 0.042$ whereas the GCA variance due to parental inbreds was $V_{GCA(1)} = 0.050$. Comparison of estimates of V_R in T-BLUP and TM-BLUP indicated that the RFLP loci did not account for any new variability. Rather, V_R and V_{SCA} remained constant between the two BLUP models and the use of markers in TM-BLUP simply accounted for a portion of the GCA variance due to parental inbreds in T-BLUP.

T-BLUP versus TM-BLUP and M-BLUP

In the B × D heterotic pattern, correlations between predicted and observed single-cross performance increased by <0.01 when TM-BLUP was used instead of T-BLUP (Table 10–4). Thus, BLUP based on trait data alone was useful for routine identification of superior single crosses prior to field testing, and the incorporation of RFLP information in the model did not lead to more effective predictions. Predictions with markers alone (M-BLUP) were less effective than with either T-BLUP or TM-BLUP, especially for yield and stalk lodging.

These results agreed with those of Knapp (1994), who found from theory that marker-only selection is never superior to selection for an index incorporat-

Table 10–3. Estimates of genetic and residual variances for yield (t ha^{-1}), moisture (g kg^{-1}), and stalk lodging (%) with T-BLUP and TM-BLUP in the B × D maize heterotic pattern.

T-BLUP†	TM-BLUP	Yield	Moisture	Stalk lodging
$V_{GCA(1)}$ ‡		0.192	193.5	2.84
	$V_{GCA(1)}$	0.050	126.9	0.32
	$V_{MGCA(1)}$	0.042	6.1	0.47
$V_{GCA(2)}$		0.165	254.1	1.67
	$V_{GCA(2)}$	0.130	122.6	0.83
	$V_{MGCA(2)}$	0.010	12.1	0.12
V_{SCA}		0.030	29.9	0.72
	V_{SCA}	0.027	29.1	0.64
V_R		1.409	291.1	18.25
	V_R	1.415	292.1	18.21

† T-BLUP, best linear unbiased prediction (BLUP) based on trait phenotypes; TM-BLUP, BLUP based on trait phenotypes and marker genotypes at RFLP loci with significant ($P \leq 0.1$) general combining ability (GCA) effects.

‡ $V_{GCA(1)}$ = GCA variance of inbreds in heterotic group B; $V_{GCA(2)}$ = GCA variance of inbreds in heterotic group D; V_{SCA} = specific combining ability variance; V_R = residual variance; $V_{MGCA(1)}$ = average marker GCA variance across RFLP loci with significant effects in heterotic group B; $V_{MGCA(2)}$ = average marker GCA variance across RFLP loci with significant effects in heterotic group D.

Table 10–4. Correlations between predicted and observed single-cross performance with T-BLUP, TM-BLUP, and M-BLUP in the B × D maize heterotic pattern.

Trait	T-BLUP†	TM-BLUP	M-BLUP
Yield	0.764 ‡	0.765	0.341
Moisture	0.934	0.934	0.690
Stalk lodging	0.641	0.647	0.311

† T-BLUP, best linear unbiased prediction (BLUP) based on trait phenotypes; TM-BLUP, BLUP based on trait phenotypes and marker genotypes at RFLP loci with significant ($P \leq 0.1$) general combining ability effects; M-BLUP, BLUP based on marker genotypes.
‡ All correlation coefficients were significantly different from zero at $P \leq 0.05$.

ing both trait and marker information. Knapp also found that selection for such marker-trait index would be superior to phenotypic selection only when heritability of the trait is low (<0.5). But in the B × D heterotic pattern, heritability was 0.62 for yield, 0.91 for moisture, and 0.63 for stalk lodging. As Knapp noted, a dilemma is that the usefulness of marker-assisted selection is greatest when trait heritability is low, yet trait heritability must be high for reliable detection of markers affecting the trait.

The marker data provided information on the covariance between single crosses in TM-BLUP and M-BLUP. The detection of RFLP loci with significant GCA effects with TM-BLUP depended on linkage disequilibrium between the marker and a quantitative trait locus (QTL). Emphasis was placed upon the estimation of the variance of marker GCA effects, rather than the variance of the marked QTL effects. The latter would have required estimation of the recombination rate (r) between the marker and the QTL. The use of TM-BLUP for estimating QTL GCA variances (V_{QGCA}) and r with single cross data is yet to be studied. If estimates of V_{QGCA} and r are available, then (i) the probability of identity by descent (Falconer, 1981) of QTL alleles can be estimated and (ii) effects of marked QTL alleles can be fitted as random effects within each parental inbred (i.e., the effect associated with a given marker allele will not be the same in inbreds i and i'). Such an approach would remove the constraint of homogeneous $V_{MGCA(1)}$ and $V_{MGCA(2)}$ across RFLP loci in the current study. Still, marker loci in linkage equilibrium with QTL will not be detected. Even with equilibrium, however, marker data do provide information on covariances of marked QTL effects (Wang et al., 1995; Bernardo, 1998) and may lead to more effective predictions of single-cross performance.

The SCA effects of paired RFLP alleles from heterotic groups B and D were not included in TM-BLUP and M-BLUP. Inclusion of marker SCA effects would have grossly inflated the size of the coefficient matrix in the mixed-model equations, causing severe computational difficulties.

Although RFLP information was not useful for choosing the best single crosses among existing inbreds, such information may be useful for choosing parents to develop new inbreds. None of the parental inbreds in this study was fixed for the favorable marker allele at all RFLP loci associated with a given trait. Selection for the most favorable marker allele at each RFLP locus may be done in an attempt to concentrate as many favorable QTL alleles as possible in a single new inbred.

REFERENCES

Bernardo, R. 1992. Retention of genetically superior lines during early-generation testcrossing of maize. Crop Sci. 32:933–937.

Bernardo, R. 1996. Best linear unbiased prediction of maize single-cross performance. Crop Sci. 36:50–56.

Bernardo, R. 1998. A model for marker-assisted selection among single crosses with multiple genetic markers. Theor. Appl. Genet. (in press).

Efron, B. 1981. The jackknife, the bootstrap and other resampling plans. Soc. Ind. Appl. Math., Philadelphia, PA.

Emik, L.O., and C.E. Terrill. 1949. Systematic procedures for calculating inbreeding coefficients. J. Hered. 40:51–55.

Falconer, D.S. 1981. Introduction to quantitative genetics. 2nd ed. Longman, London.

Hallauer, A.R. 1990. Methods used in developing maize inbreds. Maydica 35:1–16.

Hallauer, A.R., and J.B. Miranda, Fo. 1981. Quantitative genetics in maize breeding. Iowa State Univ. Press, Ames.

Henderson, C.R. 1985. Best linear unbiased prediction of nonadditive genetic merits in noninbred populations. J. Anim. Sci. 60:111–117.

Johnson, B. 1989. The probability of selecting genetically superior S_2 lines from a maize population. Maydica 34:5–14.

Knapp, S.J. 1994. Selection using molecular marker indexes. p. 1–11. *In* Analysis of molecular marker data. Proc. Joint Plant Breed. Symp. Series. Am. Soc. Hort. Sci. and Crop. Sci. Soc. Am. Corvallis, OR. 5–6 Aug. 1994. ASHS, Alexandria, VA.

Stuber, C.W., and C.C. Cockerham. 1966. Gene effects and variances in hybrid populations. Genetics 54:1279–1286.

Wang, T., R.L. Fernando, S. van der Beek, M. Grossman, and J.A.M. van Arendonk. 1995. Covariance between relatives for a marked quantitative trait locus. Genet. Sel. Evol. 27:251–274.